WARFARE
AT SEA

WARFARE
AT SEA

Motorbooks International
Publishers & Wholesalers

This edition first published in 1997 by
MBI Publishing Company, 729 Prospect Avenue,
PO Box 1, Osceola, WI 54020 0001 USA

Reprinted 1998

MBI Publishing Company books are also available at discounts in
bulk quantity for industrial or sales-promotional use. For details
write to Special Sales Manager at Motorbooks International
Wholesalers & Distributors, 729 Prospect Avenue, PO Box 1,
Osceola, WI 54020-0001 USA.

Library of Congress Cataloging-in-Publication Data Available.

ISBN 0-7603-0405-X

Editorial and design: Brown Packaging Books Ltd
Bradley's Close, 74–77 White Lion Street, London N1 9PF

Printed and bound in The Czech Republic

*Page 6: Launch of an SM-2 anti-aircraft missile from
USS Lake Erie.*

CONTENTS

In the late 1950s, the major threat perceived by the US Navy was that of saturation air and missile attack. The Soviets were expected to try to overwhelm fleet missile defences with a large number of targets which would have to be engaged simultaneously.

As existing missiles, such as Talos and Terrier, were tactically limited by the requirement that the target has to be illuminated by a dedicated radar from launch to interception, a new very long-range missile known as Typhon was proposed. This was to use new aerodynamics and electronics to overcome the saturation threat, but the project was cancelled as it became clear that it was beyond then-current technology.

Nevertheless, many of the ideas and concepts generated for Typhon were to emerge in the AEGIS system. This vastly ambitious programme was initiated in 1964, with engineering development beginning in 1969. The first AEGIS-equipped ship was the cruiser USS *Ticonderoga* (CG-47), which joined the US fleet in 1983.

The major components of the AEGIS system are the missile, the launcher, the fire-control system and the radar. The Standard SM-2 missile is semi-autonomous, using inertial navigation and terminal homing, and only needing guidance if it strays off course. This means that each missile requires much less attention from the ship's radars and fire-control computers.

AEGIS ships are now equipped with vertical launch systems, giving a rate of fire of one missile per second, and enabling them to make multiple engagements more effectively than they could when equipped with the preceding twin-rail missile launchers.

The huge octagonal 'billboards', high on the superstructure of AEGIS-

AEGIS
Fleet Defender

A Standard SM-2MR missile is fired from the forward Mk 41 vertical launch system of an AEGIS cruiser, while in the Combat Information Centre, deep within its hull, (left) crewmen are watching a detailed radar picture of all air activity within hundreds of miles.

equipped ships, are the antennae for the phased array radar system. This can detect and track over 200 targets simultaneously.

Advanced computers use threat data from the phased array radars and assign the most suitable weapon to ensure the target's destruction. AEGIS can also control the air defences of a fleet, taking target information by data link directly from the sensors of other vessels or aircraft.

Engagement sequence

1 Radar

Tracking several hundred targets at once demands radars with highly precise beams, working in three dimensions to give bearing, height and range. Unfortunately, conventional antennae rotate slowly compared with the crossing rate of high-speed aerial targets, and set high on ship masts they contribute greatly to destabilising top weight. The AN/SPY-1 phased arrays of the AEGIS system are very different. Set on the superstructure, each octagonal matrix consists of an array of some 4,100 phase-shifting elements, which are, in effect, small radar transmitters. Groups of elements can be electronically switched in turn to produce several parallel channels simultaneously, in the form of energy beams that can be shaped, swept at many times per second, and moved in any three-dimensional pattern to suit a given situation.

2 Multiple threat

Had the Cold War ever turned hot, US reinforcements crossing the Atlantic would have been a vital NATO resource. Soviet tactics called for saturation attacks on high value targets, such as carrier battle groups, leaving the convoys unprotected against further attacks. The Soviets were pioneers in the development of anti-ship missiles, and anti-ship attacks could have been launched from submarines, surface vessels, aircraft, or any combination of the three. The first missiles were large, slow and relatively easy to intercept, but later developments included supersonic terminal-dive missiles and sea-skimming missiles, which are much more difficult to knock down. In addition, Soviet attack submarines could be expected to try to get in close, making pop-up attacks from close range.

3 Air battle control

The computers at the heart of the AEGIS system manage target search, detection and tracking. They also provide target designation data for the vessel's fire-control radars. Because both missile and target are tracked, the system can tell when the missile is going off course and instruct the fire-control radar to transmit course corrections. AEGIS can also manage fleet air defences. Data from radar systems aboard other escort vessels are data-linked into the AEGIS network, and the AEGIS computers can also control the missiles fired from those vessels, assessing target priorities and selecting the most appropriate weapon to make an intercept. The ultimate aim is for AEGIS to have complete control of the air defence systems of several vessels: as yet, AEGIS cannot completely take over another warship's missile systems.

4 Firepower

The reliability and rate of fire of its launcher have a great impact on the efficiency of an air defence system as a whole. The first AEGIS cruisers were equipped with twin-rail launchers, but from the sixth vessel onwards they have been fitted with Mk 41 Vertical Launch Systems (VLSs). These need no training towards the target, do not require the complex magazine and launcher equipment of previous systems, and are easy to replenish. A Mk 41 launcher holds 61 missiles, but takes up less space than the twin-rail launcher and 44-missile magazine of the preceding Mk 26 system. AEGIS cruisers are equipped with two Mk 41 VLSs, and with a rate of fire from each launcher of one missile per second, they can engage 122 targets in a fraction over one minute. The VLS system can also fire Tomahawk cruise missiles, Harpoon anti-ship missiles and ASROC anti-submarine missiles.

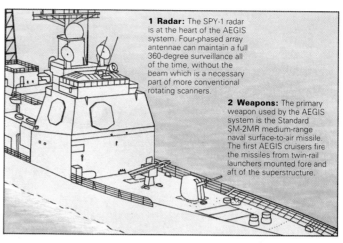

1 Radar: The SPY-1 radar is at the heart of the AEGIS system. Four-phased array antennae can maintain a full 360-degree surveillance all of the time, without the beam which is a necessary part of more conventional rotating scanners.

2 Weapons: The primary weapon used by the AEGIS system is the Standard SM-2MR medium-range naval surface-to-air missile. The first AEGIS cruisers fire the missiles from twin-rail launchers mounted fore and aft of the superstructure.

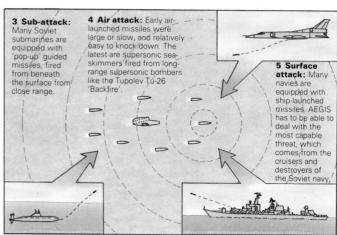

3 Sub-attack: Many Soviet submarines are equipped with 'pop-up' guided missiles, fired from beneath the surface from close range.

4 Air attack: Early air-launched missiles were large or slow, and relatively easy to knock down. The latest are supersonic sea-skimmers fired from long-range supersonic bombers like the Tupolev Tu-26 'Backfire'.

5 Surface attack: Many navies are equipped with ship-launched missiles. AEGIS has to be able to deal with the most capable threat, which comes from the cruisers and destroyers of the Soviet navy.

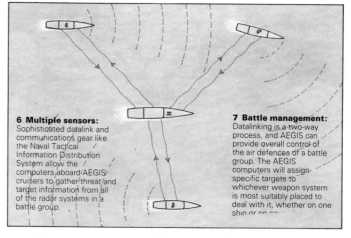

6 Multiple sensors: Sophisticated datalink and communications gear like the Naval Tactical Information Distribution System allow the computers aboard AEGIS cruisers to gather threat and target information from all of the radar systems in a battle group.

7 Battle management: Datalinking is a two-way process, and AEGIS can provide overall control of the air defences of a battle group. The AEGIS computers will assign specific targets to whichever weapon system is most suitably placed to deal with it, whether on one ship or on sea

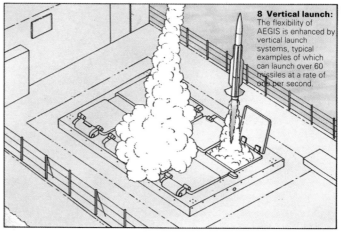

8 Vertical launch: The flexibility of AEGIS is enhanced by vertical launch systems, typical examples of which can launch over 60 missiles at a rate of one per second.

ASROC

For nearly 30 years, the US Navy's primary shipborne anti-submarine warfare system has been the weapon-carrying rocket known as ASROC. The initial planning that led to the system dates back even further, to the end of World War II.

In 1945, the US Navy began thinking about developing a weapon carrying a homing torpedo designed to engage stand-off targets at ranges of 12 kilometres or more. Although the sonar systems then available had such short ranges that they could not possibly meet the target detection requirements, considerable effort was put into designing such a weapon.

The all-weather RUR-5A ASROC anti-submarine rocket was the result of combining the features of the RAT-C rocket-thrown torpedo weapon with the concept of the rocket-thrown nuclear depth charge. ASROC was to be a ballistic rocket with a range of just over 9 kilometres, fired from a 12-round pepperpot launcher, the whole system to be paired with the newly developed SQS-23 hull sonar which had a similar direct-path detection range. The payload was to be a Mk 44 lightweight homing torpedo or a nuclear depth charge. In the event, enlargement of

the rocket meant that the launcher capacity was cut to eight rounds. ASROC was also designed to be fired from Mk 10 Terrier and Mk 26 Standard SAM launchers.

The system entered service in 1962. The Mk 44 torpedo it carried was soon replaced by the more effective Mk 46. The nuclear version of ASROC carries a 1.5-kiloton W-44 warhead in a Mk 17 nuclear depth charge. The entire nuclear system was tested under operational conditions for the first and only time in May 1962, when destroyer USS *Agerholm* (DD826) fired a nuclear-tipped ASROC during Operation Swordfish.

For the 1990s, the US Navy is developing a vertical launch ASROC for use in the increasing number of vessels with vertical launch systems. It will have longer range and will carry an improved payload comprising a Mk 46 NEARTIP or a Mk 50 Barracuda torpedo.

Because of the rapid development of Soviet submarine and submarine-launched weapon designs in the 1980s, ASROC's short range means that it is no longer a primary ASW weapon. That is now the role of the specialist ASW helicopter. Nevertheless, ASROC is still useful, both as the inner ring of a layered anti-submarine

The ASROC ASW missile can be torpedo- or nuclear-tipped. The latter creates a massive explosion, as seen in the only live test which occurred in 1963 (inset).

defence and as a weapon for use against conventional submarines. After all, combat at sea with the successors to the USSR is now highly unlikely, but operations against less capable opponents could occur almost anywhere in the world.

ASROC has been sold to a number of navies, and in addition to the United States, it is in service with Brazil, Canada, Germany, Greece, Italy, Japan, Pakistan, South Korea, Spain, Taiwan and Turkey. It has been fitted to more than 240 vessels.

Specification
RUR-5A ASROC

Type: anti-submarine weapon-carrying rocket

Dimensions: length 4.6 m; diameter 32.5 cm; span 84.5 cm

Performance: maximum speed Mach 0.8; minimum range 2 km (torpedo) or 3.5 km (nuclear depth charge); maximum range 12 km; range of torpedo payload 11 km

Launch weight: 435 kg

Warhead: one lightweight torpedo (Mk 44, Mk 46 or Mk 50) or one Mk 17 nuclear depth charge

Engagement sequence

1 Target data

The first indication of a target's presence will come from a ship's passive sonars. Modern vessels are equipped with towed arrays, which under ideal conditions can detect submarines at distances of 50 kilometres and more. Passive sonars are not very good at fixing positions, so to nail down the boat a helicopter must be sent out. If the submarine eludes the searchers and gets to within about 15 kilometres, the vessel can use its active sonar, blasting out sound from hull-mounted sonar domes. From the echoes it can fix the target's course, depth and speed with some precision. The ship's ASW fire-control system will then direct and elevate the ASROC rocket launcher to the optimum launch position.

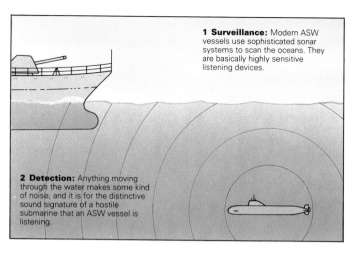

1 Surveillance: Modern ASW vessels use sophisticated sonar systems to scan the oceans. They are basically highly sensitive listening devices.

2 Detection: Anything moving through the water makes some kind of noise, and it is for the distinctive sound signature of a hostile submarine that an ASW vessel is listening.

2 Missile phase

The rocket which powers ASROC is unguided, following a ballistic trajectory. The solid fuel rocket booster is jettisoned at a predetermined point in the flight, the payload continuing on unpowered. The main purpose of ASROC is to get the weapon it carries into the general vicinity of the target in the shortest possible time. With a maximum speed of Mach 0.8, ASROC can carry a torpedo or a nuclear depth charge to a point 10 kilometres away in just over half a minute. Even if a submarine knows it has been detected, which it will do having heard the ASW vessel's sonar, it does not have much chance of manoeuvring to safety in 30 seconds.

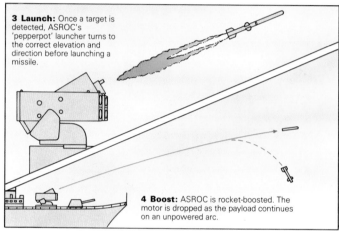

3 Launch: Once a target is detected, ASROC's 'pepperpot' launcher turns to the correct elevation and direction before launching a missile.

4 Boost: ASROC is rocket-boosted. The motor is dropped as the payload continues on an unpowered arc.

3 Torpedo phase

When ASROC carries a homing torpedo, it will deploy a parachute before the weapon hits the water, both to slow it down on impact and to ensure that the torpedo enters the water at the optimum search angle. Upon entering the water, the submarine will adopt a helical search pattern, its onboard passive sonar hunting for the target. Once it acquires the target it will attack at 50 knots, using active sonar to ensure accuracy. If it misses the target the torpedo will turn and attack again, repeating this for as long as its fuel lasts. The weapon's shaped-charge warhead is designed to punch a hole in the toughest of pressure hulls.

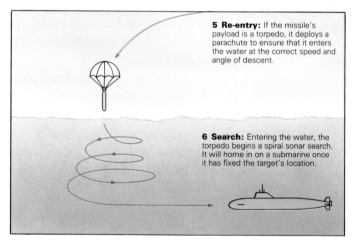

5 Re-entry: If the missile's payload is a torpedo, it deploys a parachute to ensure that it enters the water at the correct speed and angle of descent.

6 Search: Entering the water, the torpedo begins a spiral sonar search. It will home in on a submarine once it has fixed the target's location.

4 The nuclear option

The high speed and deep-diving depths of nuclear submarines mean that they might escape attacks from lightweight torpedoes. Therefore ASROC can also be fitted with a Mk 17 nuclear depth charge, although the changing nature of the submarine threat means that such weapons are no longer being taken to sea in peacetime. The Mk 17 has a 1.5-kiloton warhead, which will destroy any submarine within a kilometre and cause severe damage at two kilometres. Nuclear-tipped ASROC is not parachute-retarded. It hits the water at speed, sinking rapidly to a predetermined depth before detonating. Nuclear ASROC also has a limited anti-surface ship capability, and can be used to attack coastal targets.

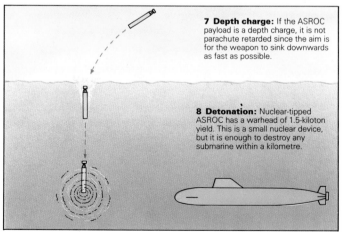

7 Depth charge: If the ASROC payload is a depth charge, it is not parachute retarded since the aim is for the weapon to sink downwards as fast as possible.

8 Detonation: Nuclear-tipped ASROC has a warhead of 1.5-kiloton yield. This is a small nuclear device, but it is enough to destroy any submarine within a kilometre.

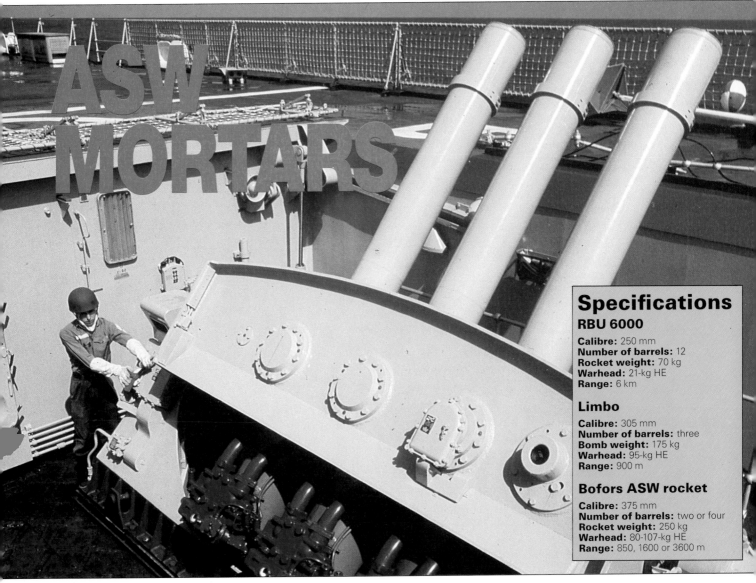

ASW MORTARS

Ever since its introduction, the submarine has been among the few weapons that could truly be called strategic. Its ability to choke the seaborne commerce of a nation has from the start made the techniques of seeking out and destroying such vessels a vital part of any navy's armoury.

Methods in use have varied greatly over the years. At one extreme, the nuclear depth bomb was awesomely unspecific, destroying everything within a kilometre or more of its detonation. At the other extreme are the high-tech 'smart' weapons, which home in on the sound generated by a submarine.

The Royal Navy's Limbo system fired 175-kilogram depth charges from this three-barrelled launcher that was stabilised for pitch and roll. Its range was less than 1000 metres.

Between the two are the conventional high-explosive depth charges. These require precise placing. Large weapons can be fitted with proximity fuses, since the force of their detonation can damage a submarine even with a near miss, but smaller weapons have to be contact-fused, since only by scoring a direct hit can their small explosive charges damage or destroy an enemy boat.

The first depth charges were simply barrel-shaped high-explosive devices, rolled off the back of a ship and detonated by hydrostatic fuses. Unfortunately, this meant that the attacking ship had to pass directly over the sub-

The Swedish navy's Bofors 375-mm anti-submarine rocket system is still in service. In the shallow waters and poor sonar conditions of the Baltic, the accuracy of homing torpedoes is greatly reduced and a hail of unguided bombs is more likely to get a result.

merged submarine, and in the last stages of the attack the ship's hydrophones would be drowned out by its own passage.

During World War II, depth charge projectors were used which could hurl their weapons to the side of the ship, allowing attacks to be made with more precision. The most effective weapons used during the war were those like Hedgehog and Squid, which could fire several small charges ahead of the vessel at once, creating a pattern of charges around the target submarine. Although largely supplanted by guided torpedoes, pattern firing weapons still have a place at sea.

The British Limbo system consists of a triple-barrelled mortar, stabilised in pitch and roll. It is controlled by the ship's ASW fire-control system and fires a pattern of three bombs to give a three-dimensional explosive burst around the target. The bombs use either a pre-programmed pressure fuse or a delayed action time fuse, which is set by remote-control before firing using data gathered from the launch ship's sonar. The maximum engagement depth is 375 metres, and the effective range is about a kilometre. Although considered obsolete, Limbo remains a potent weapon in shallow waters, where acoustic homing torpedoes are at a distinct disadvantage due to noise reflection from the sea bottom.

Soviet ASW weapons

The Soviet navy and its successors have also used large numbers of pattern firing weapons, especially multi-barrelled ASW rocket launchers designated *Raketnaya Bombometnaya Ustaovka*, or RBU. These use the ahead-firing 'Hedgehog' principle, but each projectile is powered for greater range. All of the rockets are fitted with contact or magnetic fuses.

The most widely used RBU is the RBU 6000, which entered service in 1962. Twelve launch tubes are arranged in a horseshoe shape, and fuses are set automatically, with the rockets being fired in pairs so as to straddle the target. The projectiles weigh 70 kilograms overall, and the system has an effective range of about six kilometres. The RBU 1000 is a similar but larger system, with six barrels firing 300-mm calibre rockets out to a kilometre.

Sweden's position on the shallow Baltic means that it is faced with unique ASW problems, and it too has long relied on depth charge type weapons. Bofors developed a 375-mm rocket system in the 1950s, which has appeared in two- and four-tube versions. The missiles are streamlined, ensuring a predictable and accurate underwater trajectory.

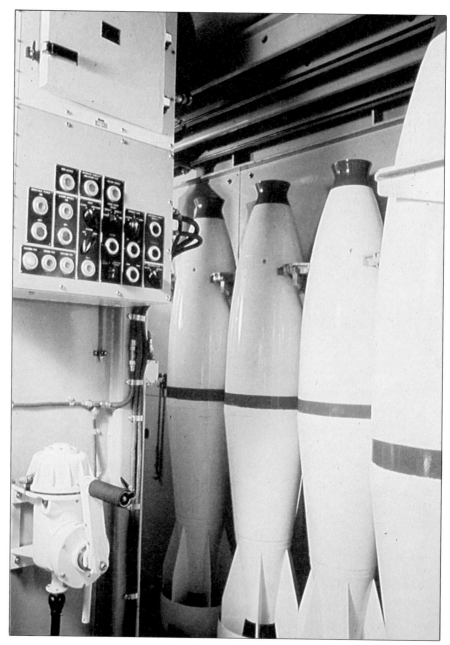

Above: The bombs fired by the Bofors ASW rocket system are streamlined for faster underwater travel. Carrying a 100-kilogram warhead, they can be fired to a maximum range of 3.6 kilometres.

Below: The Soviet navy makes extensive use of multi-barrelled ASW mortars such as these RBU 6000 weapons seen here on a 'Petya' class frigate. The 70-kilogram rockets are fired in pairs.

EXOCET!

The anti-ship missile has in recent years become the weapon with which small nations have been able to present a serious threat to the major naval powers. Argentina's success in sinking the destroyer HMS *Sheffield* and the container ship *Atlantic Conveyor* in the waters off the Falklands in 1982 brought the anti-ship missile into prominence, especially in the shape of France's best-selling Exocet.

The Exocet missile originated from a French naval requirement of the late 1960s. Developed by Aérospatiale, it proved successful in tests, with 27 out of 30 trial missiles scoring direct hits on their targets. The first customer for Exocet was the Royal Navy, which ordered 200 missiles while the system was still in the design stage and with whom the missile entered service in the mid-1970s. Since then, more than 3,000 rounds have been ordered, and Exocet is in service with some 25 navies around the world.

MM.38 missile

The basic MM.38 missile is a two-stage solid-propellant surface-to-surface weapon fitted with four cruciform wings and four tail control surfaces. It is stored in a large, rectangular packing crate-like container-launcher. Normally mounted aboard ships, the crates can be used as improvised coast-defence weapons, as the Argentines proved in the Falklands, when a shore-launched Exocet managed to severely damage the large missile-destroyer HMS *Glamorgan*.

MM.40 has an improved sustainer motor and a more compact container-launcher. Although larger and with longer range than the earlier version,

the compact mount means that more rounds can be carried in the same space occupied by the MM.38.

Although the majority of Exocets delivered have been ship-launched weapons, the first variant to see action was the air-launched AM.39. This is a shorter and lighter version of the MM.38, which incorporates a one-second ignition delay on the motor to allow the weapon to fall safely clear of the launch aircraft before firing.

Production of the AM.39 began for the French navy in 1977, and among its early export customers were Argentina and Iraq. The Iraqis were the first to use the AM.39 in anger, during the long-running Gulf war with Iran. Iraqi Exocets, launched from French-supplied Super Frelon helicopters, were used to sink at least three Iranian warships and destroyed or damaged many merchant vessels. Later in the war, an Iraqi Mirage seriously damaged the frigate USS

An MM.40 Exocet missile blasts free from its launch tube, its fold-away fins having flicked to their flight positions already. Exocet can be fitted to vessels as small as fast-attack craft.

Stark in an 'accidental' attack.

The Argentines used the AM.39 in the South Atlantic. Five Super Etendard-launched missiles, hastily brought into service with the help of French technicians, were fired. Two hit their targets, and in spite of the failure of their warheads to explode, caused enough damage to sink the *Sheffield* and the *Atlantic Conveyor*.

The final variant of the Exocet is the SM.39 submarine-launched missile. The SM.39 is carried in a capsule which can be launched through standard torpedo tubes. Like the longer-ranged American Harpoon, the missile is released from the capsule on reaching the surface of the water, launching into Exocet's normal sea-skimming flight profile.

Specification
Aérospatiale AM.39 Exocet

Type: air-launched anti-ship missile
Dimensions: length 4.7 m; diameter 35 cm; wing span 1.1 m
Weight: 652 kg
Propulsion: two-stage solid-propellant rocket
Performance: maximum speed approx 1100 km/h or Mach 0.93; range 50-70 km depending upon launch altitude
Guidance: inertial with active radar-homing in terminal phase
Warhead: 165-kg high explosive

HMS Fife fires one of the four MM.38 missiles that took the place of its original 'B' turret, with its twin 4.5-in guns. Fife is now the Chilean destroyer Blanco Encalada.

Engagement sequence

1 Surface launch

The basic MM.38 is launched from a box-like container. The longer-ranged MM.40 has folding fins, and four missiles in their compact launch tubes take up very little more space than a single MM.38. The targeting procedure is similar for both the missile variants. Before firing, the range and bearing to the target are determined by the launch platform's radar and fire-control system. Target speed and course are assessed, and the target's position by the end of the missile's flight is predicted. That data is fed into the missile's guidance system. Ships armed with the MM.38 are equipped with the ITS (*Installation de Tir Standard*), or standard fire-control system. Those firing MM.40 are fitted with an updated digital system known as ITL, or *Installation de Tir Légère*.

2 Air launch

The AM.39 air-launched Exocet is virtually the same missile as the ship-launched MM.38, but it has a smaller booster rocket. Since it is being carried by an aircraft or helicopter, it is already in the air and travelling at speed. Even with the smaller booster, the AM.39 has a longer range than the MM.38. As with the surface-launched variants, the missile is fed target data gleaned from the launch aircraft's air-to-surface radar. The missile is then launched manually by the pilot when he is confident that the target position, course and speed have been fixed. For helicopter launch the missile is fitted with a one-second ignition delay to allow the Exocet to drop safely clear of the helicopter before the booster fires.

3 Submarine launch

Since the beginning of this century, the torpedo has been the submarine's primary anti-ship weapon. However, getting close enough to mount a torpedo attack in the face of modern ASW weapons can be suicidal. The SM.39, like the American Sub-Harpoon, is designed to enable submarines to make stand-off attacks. With a range of some 40 kilometres, it can be launched against surface ship targets picked up and identified by the launch submarine's sonar systems. Sonars are not very good at estimating range, but since they are rarely effective in a direct path beyond about 20 kilometres, it means that if a target can be heard it is probably within range. Unlike the American Harpoon, the SM.39 is boosted while under water, so that it reaches cruise speeds within seconds of breaking the surface.

4 Flight profile

Once Exocet is launched towards the predicted target position, it settles into low-altitude cruise, controlled by an inertial navigation system and radio altimeter. Cruise height can be set low enough to escape detection from the target's radars, but high enough to give its own active radar a chance to seek out the target. At approximately 10 kilometres away from the target's predicted position, the missile activates its own radar, which begins to scan for the target. Once acquired, the missile drops even lower towards the sea as it homes in. Just before impact it descends again to within three metres of the surface to ensure a hit as close to the waterline as possible. A contact-delayed fuse ensures that the 165-kilogram shaped-charge warhead does not explode until it is deep inside the target.

1 Surveillance: Initial target indications come from the launch platform's radar. Exocet was designed with a range of 40 kilometres, which matches the radar horizon of a medium-sized warship.

2 Launch: Exocet launchers are elevated to 12 degrees. A two-second boost gets the missile up to cruise speed, and it never climbs more than 70 metres over the water.

3 Air-to-surface: Air-launched Exocet can be launched from longer ranges than surface variants, since an aircraft's detection systems are higher up and have longer radar horizons than those of a ship.

4 Air launch: The Exocet AM.39's is shorter and lighter than the ship-launched missile, and when launched from helicopters has a one-second ignition delay to ensure that the missile's exhaust is well clear of the launch platform.

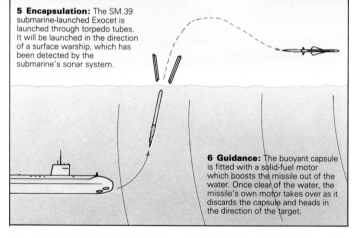

5 Encapsulation: The SM.39 submarine-launched Exocet is launched through torpedo tubes. It will be launched in the direction of a surface warship, which has been detected by the submarine's sonar system.

6 Guidance: The buoyant capsule is fitted with a solid-fuel motor which boosts the missile out of the water. Once clear of the water, the missile's own motor takes over as it discards the capsule and heads in the direction of the target.

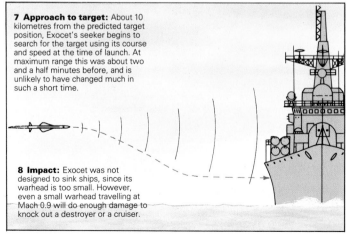

7 Approach to target: About 10 kilometres from the predicted target position, Exocet's seeker begins to search for the target using its course and speed at the time of launch. At maximum range this was about two and a half minutes before, and is unlikely to have changed much in such a short time.

8 Impact: Exocet was not designed to sink ships, since its warhead is too small. However, even a small warhead travelling at Mach 0.9 will do enough damage to knock out a destroyer or a cruiser.

GUNS OF THE BATTLESHIP

The mighty 16-in guns of the US Navy's 'Iowa' class battleships have supported US ground forces in Korea, Kuwait, Lebanon and Vietnam.

The 1224-kilogram shells were originally designed to smash armoured warships during World War II.

The gun-armed battleship dominated war at sea for over 400 years, but it seemed obsolete by the end of World War II. Between them, the aircraft carrier and the submarine had ousted it from pride of place in the navies of the world. However, half a century later the battleship was back at the centre of things.

The US Navy in the late 1970s knew it could not afford all the carrier battle groups it needed to project US power around the globe. As an alternative, Navy planners looked at the four mothballed 'Iowa' class battleships. Well preserved, having seen little use since 1945, the 'Iowas' were designed to survive a fight with the enormous guns of the Japanese super battleship *Yamato*. The relatively puny challenge of modern anti-ship missiles would hardly even dent the foot-thick armour protecting their vitals. Even in middle age, powerful steam turbines drive the 57,000-tonne vessels faster than most of the rest of the US Navy.

The broad decks of the 'Iowas' had room for plenty of advanced modern weaponry, including Tomahawk cruise missiles and Harpoon anti-ship missiles. Four Phalanx CIWS could provide point defence against close-flying missiles. Upgraded electronics and combat systems suitable for warfare in the 1990s would have to be fitted. Each could be reactivated and modernised for the cost of building a single frigate. To critics, the prospect

of reinvesting in battleships was a triumph of nostalgia over common sense. But others, including the Reagan administration, thought otherwise, and the battleship returned to duty.

Even if missiles were all the armament *Iowa* and her sisters carried, they would still be warships to be reckoned with, but their guns are the icing on the cake. A dozen 5-in guns in the secondary armament are impressive enough, but it is the battery of nine Mk 7 16-in guns which is the knockout.

Most powerful gun

The Mk 7 is one of the most powerful pieces of artillery ever built. Each gun weighs in at 108 tonnes, and an entire triple turret comes to 1730 tonnes, or as much as a small frigate. Each gun can hurl a 1.2-tonne projectile nearly 40 kilometres, and a battleship's magazines hold more than 1,200 high-explosive and armour-piercing shells. In action against modern, thin-skinned warships, such firepower could devastate fleet after fleet without the need for replenishment. Against shore targets, the battleship's ability to blanket an entire area means that enemy installations can safely be knocked out before the marines arrive.

A memory of former days they might be, but the reborn 'Iowas' are among the world's most powerful

warships. They have made show-the-flag visits around the world. *New Jersey*'s huge guns have spoken in anger off the Lebanon, and *Missouri* and *Wisconsin* used guns and missiles in the Gulf.

The Gulf War proved to be the battleship's swan song, at least for the moment. Battleships are large vessels, with large crews, and they are far from cheap to run. In these days of peace dividends and defence cuts that is a big handicap, so in spite of capabilities unmatched by any other warship, the 'Iowas' are going back into mothballs.

Specification

Mk 7 16-in gun

Length: 20.3 metres (50 calibres)
Weight: 108 tonnes
Elevation: -2°/+45°
Projectile weight: 1224 kg
Turret weight: 1730 tonnes
Range: 39000 m
Barrel life: 300 full-charge rounds

Engagement sequence

1 Turret

The 16-in guns and lightweight turrets used aboard the 'Iowa' class were a product of the inter-war Washington Treaty, which was an agreement limiting the sizes of battleships and their guns in the hope of avoiding a naval arms race. The Mk 7 gun was much lighter than preceding 16-in weapons, although it was also more powerful. It was the largest gun and turret combination that a battleship built to the treaty limit of 45,000 imperial tons could carry. As with all battleship turrets, the walls are heavily armoured to protect the highly explosive propellant and the big gun's 1.2-tonne projectiles.

2 Powder

The most volatile part of a battleship's weapon system is the projectile powder used to blast the projectiles over great distances. Powder is stored in measured amounts in cotton and silk bags, which are hoisted from the magazines direct to the guns when needed. The magazines are positioned well below the waterline, for maximum safety in case of accidental explosion. Yet, even those precautions would come to nought in an explosion, as the many tons of propellant would be enough to smash the big vessel to splinters if they ignited accidentally.

1 Powder
The most volatile part of a battleship's ammunition is the powder used to blast the shells over great distances. Powder is stored in cotton and silk bags in magazines well below the waterline for maximum safety, and is hoisted direct to the turret when required

2 Projectiles
Projectiles for the 16-in guns are stored behind thick armour on two levels of the barbette. Currently, the battleships are using up the stock of World War II and Korean War vintage projectiles, most of which are high capacity HE shells.

3 Handling
Shells and powder come to the turret on separate hoists. Surprisingly, the powder comes up a single-stage hoist. The doors at the top and bottom of the barbette, which can not be opened simultaneously, ensure no contact between turret and powder store in the event of an explosion.

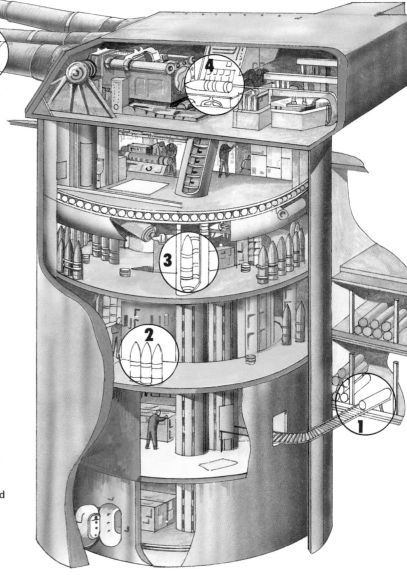

3 Projectiles

Projectiles for the 16-in guns are stored behind thick armour on two levels of the barbette, or turret. In the days when battleship might have been expected to fight battleship, many of the shells were armour-piercing, but those in use in the 1980s and 1990s have been Korean War-vintage high-explosive. Shells and powder come up to the turret on separate hoists. The heavy armour-plate doors at the top and bottom of the barbette cannot be opened simultaneously, thus ensuring that there is no contact between the turret and the magazines in the event of an explosion.

4 Load and fire

Shells are united with bags of propellant only in the breech of the gun. Six bags of propellant are used to fire one projectile. The charge can be varied, however, reducing range while at the same time reducing barrel wear. Reduced charges use bags of the same thickness as full charges, but they are of smaller diameter. Detonation of the propellant creates 17 tonnes of pressure in the breech, which is enough to drive a 1.2-tonne projectile with a muzzle velocity of more than 800 metres per second. This, with the gun at its maximum elevation of 45 degrees, results in the shell being lobbed more than 38 kilometres down range.

4 Loading
Shells are united with six bags of propellant in the breech. Reduced charges use bags of the same thickness but of smaller diameter. The lighter charge cuts pressure in the breech on firing, reducing gun wear and increasing barrel life.

5 Firing
The shells are propelled down the barrel by an explosion which creates 17 tonnes of pressure in the breech. By the time it leaves the muzzle, a 1.2-tonne shell is travelling at 800 m/sec. Barrels need replacing after firing 300 rounds.

HARPOON

Harpoon is the Western world's most widely used anti-ship missile. Here it blasts off on its 120-kilometre journey to an enemy warship.

A camera on board a target ship captures the Harpoon picoseconds before impact. The missile flies very low, but pops up at the last moment to plunge into the target's upper deck.

Specification

McDonnell Douglas AGM-84A/RGM-84A Harpoon

Type: air-, ship- or submarine-launched anti-ship missile
Dimensions: length 4.58 m for RGM-84A and Sub-Harpoon, 3.84 m for AGM-84A; diameter 34.3 cm; span 91.4 cm
Weights: 667 kg for RGM-84A and Sub-Harpoon, 522 kg for AGM-84A
Performance: maximum speed Mach 0.85 (c. 1000 km/h); range up to 120 km
Warhead: 227-kg high explosive with time-delayed contact fuse
Guidance: active radar-homing

The US Navy began to show serious interest in anti-ship missiles in 1967, after the sinking of an Israeli destroyer by SS-N-2 'Styx' missiles during the Six Day War. Harpoon began life in 1967 with the requirement for a 92.5-kilometre' range anti-ship missile, designated AGM-84A, suitable for air-launch. By 1970 the requirement had been extended to include ship-launch (as the RGM-84A), and in 1972 the programme to develop the underwater-launched Sub-Harpoon was initiated.

Harpoon was designed as a simple, low-risk programme. It is a sea-skimming, active radar-homing missile designed to attack manoeuvring targets at ranges of 120 kilometres or more. One Harpoon is reckoned to be enough to destroy a fast missile craft, while two will disable a frigate. Knocking out larger targets like missile cruisers or carriers might need four or five missiles.

The basic Harpoon has been subject to regular upgrades, designed to maintain and enhance its lethality in the face of evolving naval anti-missile systems.

Within the US Navy it has been deployed on S-3, P-3, A-6, A-7 and F/A-18 aircraft, and on surface vessels from hydrofoils up through frigates,

destroyers, cruisers and battleships. Sub-Harpoon can be fired through standard torpedo tubes. The US Air Force uses B-52G heavy bombers to carry up to 12 Harpoons apiece on a sea control mission.

Britain uses all three Harpoon variants, and other users or potential operators include Australia, Canada, Denmark, Germany, Greece, Iran, Israel, Japan, Netherlands, New Zealand, Saudi Arabia, Singapore, South Korea, Spain, Thailand and Turkey.

Harpoon's combat debut of both ship- and air-launched versions

occurred during the confrontation with Libya in March 1986, when two 'Nanuchka' class missile corvettes and two 'Combattante II' fast attack craft were sunk or severely damaged.

No anti-ship Harpoons were fired during the Gulf War, but seven examples of the air-launched AGM-84E Stand-off Land Attack Missile, or SLAM, were used. SLAM is equipped with the imaging infra-red guidance system of the Maverick missile, and can be used to make all-weather precision attacks on land targets.

Engagement sequence

1 Target

Harpoon is designed to engage manoeuvring targets at ranges out to 120 kilometres or more, and newer generation missiles have even greater ranges. Initial target information can be provided by the launch platform's own radar, although this can cause problems at maximum range because the enemy will be beneath the radar horizon, in which case some kind of third-party targeting may be necessary. This is unlikely to be needed when Harpoon is air-launched at altitude, however. When the Harpoons are being fired by one of the vessels in a US Navy carrier battle group, the Grumman E-2C Hawkeye airborne early-warning aircraft will be able to give enemy surface ship positions while they are still several hundred kilometres away.

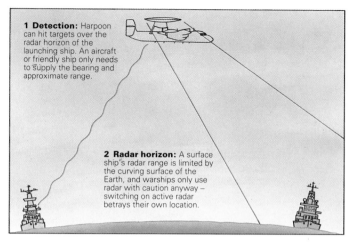

1 Detection: Harpoon can hit targets over the radar horizon of the launching ship. An aircraft or friendly ship only needs to supply the bearing and approximate range.

2 Radar horizon: A surface ship's radar range is limited by the curving surface of the Earth, and warships only use radar with caution anyway – switching on active radar betrays their own location.

2 Launch

The ship-launched RGM-84A Harpoon is boosted into flight by a small solid-propellant rocket, which is quickly exhausted and falls away. Once in the air and travelling at speed, propulsion is provided by a Teledyne-Ryan CAE J402 turbojet, which sustains the missile at about 1000 km/h for some 10 minutes of flight time. Sub-Harpoon is similar. It is stored in a capsule which is launched through a submarine's torpedo tubes. The capsule bobs to the surface, and the booster rocket fires in the same way as the ship-launched variant. The air-launched AGM-84A needs no booster, since it is already travelling through the air at several hundred kilometres per hour beneath its parent aircraft.

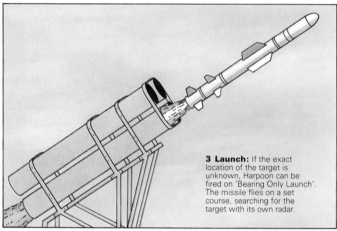

3 Launch: If the exact location of the target is unknown, Harpoon can be fired on 'Bearing Only Launch'. The missile flies on a set course, searching for the target with its own radar.

3 Profile

Harpoon is usually launched in RBL, or Range and Bearing Launch mode. It is launched under inertial guidance to the general target area, before turning on its own Texas Instruments radar at the last moment. The radar can be set for large, medium or small acquisition windows. The smaller the window, the more precise the initial target data must be, but the less time there is for the missile to be defeated by enemy counter-measures. Bearing Only Launch, or BOL mode, involves activating the radar early in flight, and scanning 45 degrees to each side of the missile's line of advance to search for a target. If no target is acquired after a pre-set time, the missile can be switched to a pre-set search pattern.

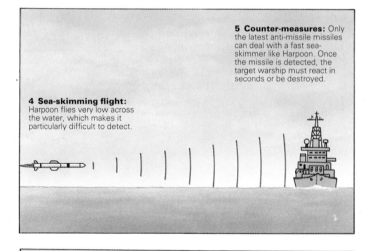

5 Counter-measures: Only the latest anti-missile missiles can deal with a fast sea-skimmer like Harpoon. Once the missile is detected, the target warship must react in seconds or be destroyed.

4 Sea-skimming flight: Harpoon flies very low across the water, which makes it particularly difficult to detect.

4 Terminal homing

Whichever guidance mode is chosen for Harpoon, once the target is detected the seeker automatically locks on. Early Harpoons skimmed the surface of the sea during flight. Once they approached the target and the homing radar had locked on, they climbed rapidly in a 'pop-up' manoeuvre before making a terminal dive attack. Current models have increased range, and follow a sea-skimming path all the way into the side of the target. Harpoon's 227-kilogram high-explosive warhead is fitted with a time-delayed contact fuse, allowing the missile to penetrate right inside the target before detonating, creating maximum damage and possible secondary fires and explosions. Modern warships are not armoured, so the missile is easily capable of penetrating their thin sides.

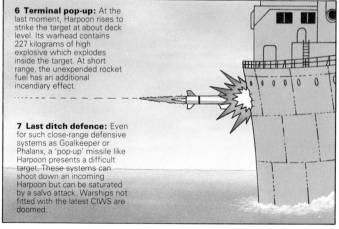

6 Terminal pop-up: At the last moment, Harpoon rises to strike the target at about deck level. Its warhead contains 227 kilograms of high explosive which explodes inside the target. At short range, the unexpended rocket fuel has an additional incendiary effect.

7 Last ditch defence: Even for such close-range defensive systems as Goalkeeper or Phalanx, a 'pop-up' missile like Harpoon presents a difficult target. These systems can shoot down an incoming Harpoon but can be saturated by a salvo attack. Warships not fitted with the latest CIWS are doomed.

HMS *FEARLESS*

HMS Fearless *was critical to the success of the Falklands campaign. It was lucky that the government's plan to scrap her had not yet been implemented when Argentina invaded the islands.*

Amphibious Assault Ship

HMS Fearless *is not a particularly attractive ship. Being an assault support ship she has no particular reason to have attractive lines, but the angular appearance of the vessel's high forward superstructure and the long, low helicopter landing deck are highly functional.* The blunt stern hides the fact that she has an internal landing craft 'hangar', and the relatively broad hull hides the large numbers of stores, accommodation decks, workshops, offices and communications centres that an assault ship requires in action. Apart from all these facilities *Fearless* has to find room for her ship's complement of 580 men,

and additional volume has to be found for her helicopters, aircrew and maintainers, to say nothing of the 380-400 Royal Marines who regard the ship as home.

Fearless was completed in 1965 and spent many years providing amphibious operations support in all parts of the world, but increasingly in Scandinavian waters as the years rolled by. However, the operating expense of *Fearless* grew even higher as general costs rose and items of equipment started to wear out, until in 1981 she was earmarked for laying off in 1984; but strings were pulled in high places and *Fearless* was reprieved. Thus in early 1982 *Fearless* was tied up at her

usual wharf in Portsmouth Dockyard undergoing an extensive refit and refurbishing programme on all parts of the ship.

Events in the South Atlantic changed all that overnight. The dockyard 'maties' worked immensely long hours to put *Fearless* back into commission, and they achieved virtual miracles as the ship was due to sail on Tuesday 6 April. As it was, the ship sailed out of Portsmouth on only one boiler as the other was still recovering from extensive repairs. But to the watching public all appeared well as they gave the small convoy a rousing send-off into the wet mist of the Solent.

At that stage *Fearless* was still relatively empty apart from the usual mountain of stores packed into all parts of the ship, including the gangways. On board were Commodore Michael Clapp (Commodore Amphibious Warfare), his staff and elements of No. 3 Commando Brigade (including the headquarters), three Westland Sea King HC.Mk 4s from No. 846 Naval Air Squadron and three Westland Scout AH.Mk 1s of No. 3 Commando Brigade. As *Fearless* moved south to Ascension Island the ship's company gradually sorted out the state of the ship and tried to establish where everything had been stacked. The commandos carried out as much training as they could, and life was busy as Ascension loomed on the horizon.

Ascension Island

Fearless arrived at Ascension on 17 April, and on the very next day there began a series of exercises to train the landing craft crews and their charges, and a long series of shifting about of men and stores from one vessel to another, nearly all the work being carried out by helicopter. To add to this, the crew of *Fearless* began a large number of operations to provide technical and other assistance to other ships in the force that was gradually assembling. (HMS *Intrepid*, the sister ship of *Fearless*, arrived on 15 April after an even more involved dockyard operation than that needed by *Fearless*). Technicians from *Fearless* were transported around the fleet carrying out all manner of tasks from electrical repairs to removing the deck sidewalls from a chartered Ro-Ro (roll-on/roll-off) ferry to convert it into a helicopter landing-decked stores vessel.

On 7 May *Fearless* and a small fleet of support vessels left Ascension and moved south towards the Falklands.

Among the many ships moving south was the *Canberra*, soon to be immortalised as 'the Great White Whale', and from her *Fearless* took on board the 600 or more men of No. 40 Commando plus all their kit and another heap of stores and equipment.

This brought the onboard total to well over 1,500 men, who had to sleep and live where and how they could.

That was on 19 May, by which time *Fearless* was in good company. Sailing with her were the *Canberra*, the ill-fated *Atlantic Conveyor*, the *Europic Ferry*, the *Elk*, the RFA *Stromness*, the *Intrepid*, and as escorts there were HMS *Ardent* and HMS *Argonaut*. *Fearless* was one of the central vessels of this small fleet, for it was on board that most of the expedition's commanders were based. It was on board *Fearless* that the assault commanders learned of their projected landing beach at San Carlos (that was on 10 May). The landing operations (Operation 'Sutton') then had to be prepared, but as they did so the weather got worse as the 'Roaring Forties' were passed; then on 20 May *Fearless* entered the Total Exclusion Zone (TEZ) around the Falkland Islands. By then the force around *Fearless* and *Intrepid* had increased: the *Canberra*, *Europic Ferry*, *Elk* and *Stromness* were still there, but to

HMS *Fearless*

HMS *Fearless* as she appeared during the early stages of the Falklands campaign. One of her four LCUs is alongside: these can carry 70 tons of stores, 250 soldiers, or a tank. The smaller LCVPs, shown hoisted in, can carry 35 soldiers or two Land Rovers. The helicopter deck is occupied by a single Wessex, but it can accommodate up to four Westland Sea Kings.

Open stern
The stern well is flooded to enable the LCUs to move in and out with their cargo. These 80-ton boats are capable of nine knots and are operated by the Royal Marines.

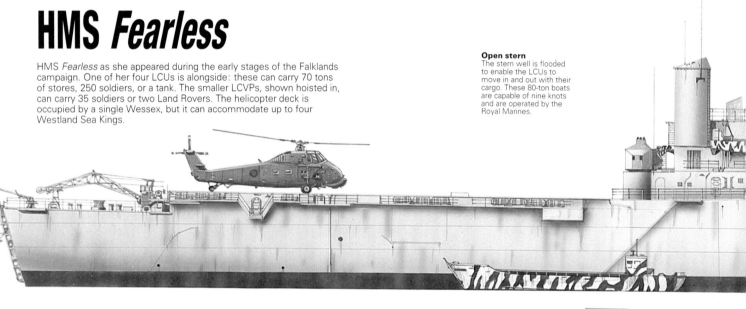

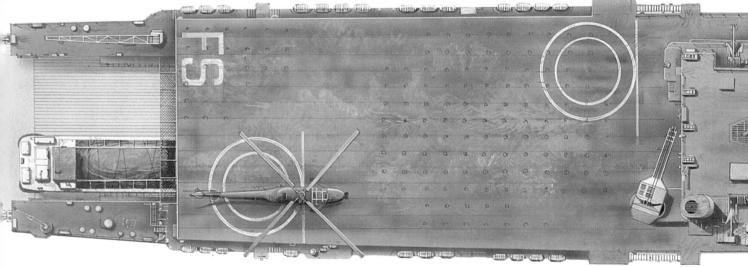

these had been added the RFA's *Sir Percival, Sir Geraint, Sir Galahad, Sir Lancelot, Sir Tristram* and *Fort Austin*. Escorts now included HMS *Broadsword, Brilliant, Ardent, Argonaut, Antrim, Plymouth* and *Yarmouth*.

As this force approached San Carlos Water it was spotted by a lone Argentinian air force BAC Canberra, but by then it was too late. *Fearless* started her war work with a vengeance: using the four landing craft carried in the stern, No. 40 Commando started to move ashore. The landing craft car-ried out the first full-scale amphibious landing in earnest for many years on the night of 21 May, and all went well.

Argentine attacks

The very next day the air attacks started. *Fearless* was at the centre of it all as Argentinian air force Dassault Mirages and McDonnell Douglas Sky-hawks swung into action. *Fearless* was an obvious target, but somehow she managed to escape the worst. Bombs fell close by, but the ship's defences of two 40-mm L/60 Bofors guns (dating from World War II) and four Sea Cat missile-launchers added their bit to the wall of defensive fire that greeted the attackers. Coming in low and relatively straight, the Argentinians presented good, if fast, targets and they suffered accordingly. The first day's tally was 17 Argentinian air force aircraft shot down, and from

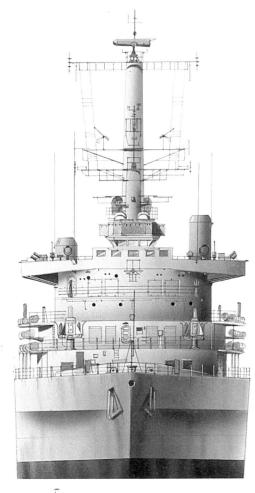

Air defence
The four yellow-topped Sea Cat missile launchers can be seen here – two forward and one either side of the superstructure. The latter pair have now been replaced by Phalanx 20-mm guns, and BMARC 30-mm weapons have replaced the obsolete Bofors guns.

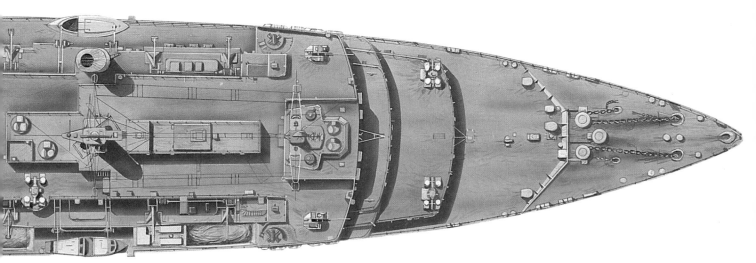

then on San Carlos Water became known as 'Death Alley' to the attackers, but they kept coming back with a degree of courage which defies description. On the debit side for the British was the loss of HMS *Ardent*, so for the British San Carlos Water became 'Bomb Alley'.

There were no air attacks on 22 May. *Fearless* concentrated on getting men and stores ashore to consolidate the beach-head. Her landing craft fussed around the Water carrying men and supplies from ship to ship and working for as long as the light held. The next day was another matter. The Argentinian air force returned in strength; this time the attackers hit and destroyed HMS *Antelope*. Before the spectacular explosion that made such photographic history took place, the little landing craft from *Fearless* were able to take off nearly all her crew and that night *Fearless* provided a place to rest for the homeless 'Antelopes'.

On 24 May the air attacks continued and *Fearless* received three casualties, the only ones of her whole campaign; three men on a Bofors gun were wounded by shrapnel and one had to be taken ashore for treatment. Below decks things were as busy as ever, the men involved having little or no idea of what was going on above decks. They had to keep working in the engine rooms, communications centres, landing craft well, stores and all the many other hidden workplaces, while above them bombs dropped and guns thumped away.

It was on 24 May that *Fearless* really started her 'mother' role in earnest. From the start of the campaign the ship had been loaning technicians and manpower to other ships, but now *Fearless* started to send battle damage parties to stricken ships. The LSL *Sir Lancelot* was struck by two unexploded bombs (UXB) and the crew had to be removed. As they left parties from *Fearless* moved in. Electrical cables were replaced, temporary repairs brought back lighting and power, new galley equipment was fitted and once the bombs had been rendered safe the crew returned. This was typical of other *Fearless* support operations after bomb damage; a party from *Fearless* carried out extensive repairs to *Argonaut* after a raid; during June a further *Fearless* party assisted in putting out fires on HMS *Plymouth* and clearing a path to unexploded bombs; repair work was carried out on the *Nordic Ferry* and the tug *Irishman*; and so it went on.

Perhaps the saddest task that men from *Fearless* had to perform was on 11 June. Using a Boeing Vertol Chinook helicopter this party had to survey the wreck of the *Sir Tristram*, damaged severely during the Bluff Cove operation. Despite the severe damage from fire the vessel was repaired to the extent that it could be used for much-needed troop accommodation once the islands had been repossessed.

Fearless did not remain in San Carlos Water throughout all the air attacks. On 28 May she was out at sea ready to take on Major General Moore and his staff along with Brigadier Wilson (commander of the 5th Infantry Brigade, the main army force for the final land operations). They were all back in 'Bomb Alley' next day, and there *Fearless* remained until 6 June. On that day *Fearless*' manpower strength again went over the 1,500 mark as Welsh Guards came aboard ready for the move to Elephant Island, where the landing craft once more put the Guards ashore.

Returning home

On 25 June *Fearless* left the Falklands and sailed north. It had not been a bad effort from a ship that had once been almost assigned to the scrapheap. For long and dangerous days she had been in the thick of it all, with bombs falling everywhere around her. Her gunners had scored some kills and many men had been accommodated in her metal interior. Her helicopters had kept men and supplies flowing in all directions. Her damage parties had kept ships afloat and her engineering bays had churned out parts for all manner of machines.

Without *Fearless* the Falkland Islands operation would probably not have been possible – not a bad thing for a ship and her company to look back on.

IKARA

Design and production of the all-weather Ikara anti-submarine missile was initially undertaken by the Australian Department of Defence. Similar in concept, though differing in design from the American ASROC, the Ikara system is meant to operate from surface ships in all weathers, and is effective out to the maximum range of most ship-borne sonars.

The concept behind Ikara is simple. It is designed to carry a lightweight homing torpedo out to the general vicinity of an underwater target. By using a guided missile as the carrier, Ikara-armed vessels can attack an enemy submarine much faster and at much greater range than ships armed with conventional torpedoes.

Ikara is powered by a solid-fuel combined booster and sustainer rocket. In all its variants it is launched on a bearing to the target, and when it reaches its attack point it drops its payload, a parachute-retarded lightweight torpedo.

The attack point is calculated from the launch platform's own sonar, or by a remote data-linked source, such as a helicopter with dipping sonar or another vessel. The information re-

ceived is fed into the launch vessel's fire-control system, which continually calculates the optimum drop position.

After launch, the fire-control system passes the revised drop position data to the missile in flight in the form of control commands, via the ship's radar/radio weapons guidance system.

Above: An Ikara roars into the sky on its way to deliver a homing torpedo to the estimated position of an enemy submarine. Introduced in the 1970s, it was a key ASW system that allowed surface ships to attack submarines at much longer ranges.

Below: The 'Leander' class frigate HMS Arethusa was commissioned in 1965 and was converted to carry Ikara in 1977, the last of six 'Leanders' to be fitted with the weapon. Arethusa was deleted in 1991.

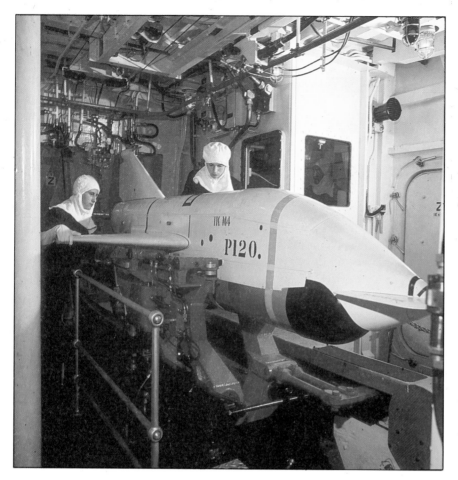

Sailors prepare an Ikara missile for launch. They are wearing anti-flash hoods to help reduce burns injuries if the ship is hit.

in the magazine and loading systems.

The system became operational aboard a variety of vessels. In Australian service, it has equipped the six 'River' class frigates and three 'Perth' class destroyers. British vessels fitted with Ikara include HMS *Bristol* and six 'Leander' Batch 1 frigates. For operational reasons, however, the Royal Navy prefers the flexibility of ship-borne helicopters, and Ikara is no longer in use. All Ikara-armed vessels have been decommissioned or are up for disposal. One ex-British 'Leander' has been commissioned into the Royal New Zealand Navy.

Brazilian navy

Ikara has been sold to the Brazilian navy, where it is known as Branik. Mounted on ASW variants of the 'Niteroi' class frigate, it differs from the British and Australian systems in that it employs a special-purpose missile tracking and guidance unit developed by Ferranti, which is fully integrated into the two fire-control systems carried by each 'Niteroi'.

Developments of Ikara include the British Aerospace Super Ikara, with a range of 100 kilometres, and the Australian-Italian Modified Ikara, which is a folding winged variant fired from a box launcher and using the guidance system of the Otomat surface-to-surface missile. All improved varieties of Ikara are designed to accept American, British, Swedish or Italian lightweight torpedoes.

Once the missile arrives at the target area, the torpedo's payload is command ejected via the same communications link. The Ikara body then continues on, clearing the area before crashing, while the torpedo descends by parachute and enters the water already oriented to commence its search pattern. The payload is usually an American Mk 44 or Mk 46 torpedo.

Since most of the mission profile takes place over the water, a submerged submarine will have no idea that it is under attack until it hears the torpedo's propulsion system or its active sonar guidance system. How-

ever, if the Ikara system is working correctly, the torpedo should be dropped right on top of the enemy boat, and the boat will have little chance of escaping.

Ikara achieved initial operational capability in the early 1970s. By this time, the British Royal Navy had expressed an interest in the Australian system, and British Aerospace took out a licence to produce an improved version of the system known as the GWS Mk 40. The main differences between the British and Australian versions of the weapon are in the interface with the fire-control systems and

HMS Ajax meets a Soviet 'Krivak' class frigate in the Atlantic. The 'Krivak' is also armed with a long-range anti-submarine missile, the SS-N-14 'Silex'.

Specification
Ikara

Dimensions: length 3.42 m; height 1.57 m; wing span 1. 52 m
Weights: missile 300 kg; payload between 200 and 250 kg
Propulsion: dual-thrust solid-propellant rocket
Performance: maximum speed c. 750 km/h; range 25 km
Guidance: command guided
Warhead: lightweight active/passive sonar homing torpedoes

'KRIVAK' IN ACTION

'Krivak' class frigates were built by the Soviet navy to hunt and sink submarines. NATO submarines certainly planned to enter Soviet waters in time of war, laying mines and intercepting Soviet SSBNs.

Designed to hunt and kill nuclear submarines in the deep waters of the world's oceans, the 'Krivak' class of frigate is faster and much more heavily armed than Western contemporaries.

When the first 'Krivak I' *bolshoy protivolodochnyy korabl'* (large anti-submarine ship) entered service with the Soviet navy during 1970, it made an immediate impression with Western naval powers, who generally saw the design in a highly impressive light, especially when compared with contemporary Western missile destroyer and frigate designs. Despite the Soviet description of ASW ship, Western observers mistook the type's actual role to be surface strike, with anti-air and anti-submarine weapons embarked only for self-protection. The reason for this misassessment was that the class shipped in a forward position a quadruple container-launcher missile assembly for the same supersonic derivative of the SS-N-2 'Styx' as supposedly carried by the contemporary 'Kresta II' class

cruisers and initially allocated the NATO designation SS-N-10. The mistake was finally resolved during the 1970s by Western naval intelligence agencies: the revised designation SS-N-14 'Silex' was adopted as the original and the incorrect SS-N-10 label was simultaneously dropped. One apparent reason for all the confusion was reportedly the result of the Soviets actually sending the 'Kresta II' class to sea without any missiles aboard for several years as the system was undergoing final testing, a brilliant piece of deception which led intelligence services to the wrong conclusions.

Secret missile

No confirmed picture of an SS-N-14 has ever been released to the technical press, but reliable reports indicate that it is in fact a solid-propellant rocket-powered winged cruise missile that conceptually resembles the Australian Ikara ASW missile in dropping a parachute-retarded acoustic-homing torpedo. The maximum and mini-

The 'Krivak I' class frigate Druzhny off Norway observes a NATO exercise during 1986. Note the twin 76-mm gun turrets; 'Krivak IIs' have single 100-mm turrets instead.

mum ranges are 55.5 kilometres and 7.4 kilometres, respectively. The maximum range figure means that to ensure the missile's terminal accuracy dependence is placed not on the ship's own sensors but on a third-

Aboard the 'Krivak II'

The 11 'Krivak II' frigates were built from 1978-81. The main difference between them and the 21 'Krivak Is' is the fitting of single 100-mm gun turrets in place of the twin 76-mm weapons mounted on the earlier ships. Originally classified as anti-submarine ships, the Soviets redesignated the 'Krivaks' as patrol or escort ships in 1978. They have a well-balanced combination of anti-surface, anti-air and anti-submarine weapons.

Midships fire-control radars
The radar systems forward of the funnel control the aft SA-N-4 'Gecko' surface-to-air missile launcher. This is below deck abaft the funnel and elevates before firing. It is visible on the deck plan below as a white circle forward of the 100-mm guns.

100-mm guns
These automatic weapons can fire up to 80 rounds per minute and have a maximum range of 15000 metres. Each shell weighs 16 kilograms.

party source. This is a capability which fits in well with known Soviet ASW practices: for long-range targets the 'Krivak' is supplied with target position data from other surface ships or airborne platforms via a data-link system after missile launch and early control by the vessel's own 'Eye Bowl' missile tracking and command-guidance link radars. The presence of a pair of these radars aboard a 'Krivak' suggests that a salvo of two SS-N-14 missiles can be used during an engagement to increase the kill probability.

To aid in the high-speed manoeuvring associated with some aspects of ASW operations the class is fitted with a four-engine gas turbine COGAG arrangement, the two smaller units being coupled up with the two larger units for full-power operations. The exaggerated overhang of the bow is a good indication of a bow-mounted sonar, probably of the medium-frequency type. For operations against submarines operating below the thermal layer, a variable-depth sonar system is located aft, inside a prominent housing on the quarterdeck. The model carried is the same low-frequency set that is fitted to most contemporary Soviet ship designs, and may be used to target the 'Silex' missiles in close-range attacks.

Improved design

The 'Krivak I' was built in tandem with the 'Krivak II' during the mid-1970s and early 1980s. The latter variant differs in having the original twin 76-mm gun turrets aft replaced

For close-range anti-submarine action the 'Krivaks' carry two 12-tube RBU 6000 anti-submarine mortars, seen here just forward of the bridge. These fire 31-kilogram bombs to a maximum range of 6000 metres.

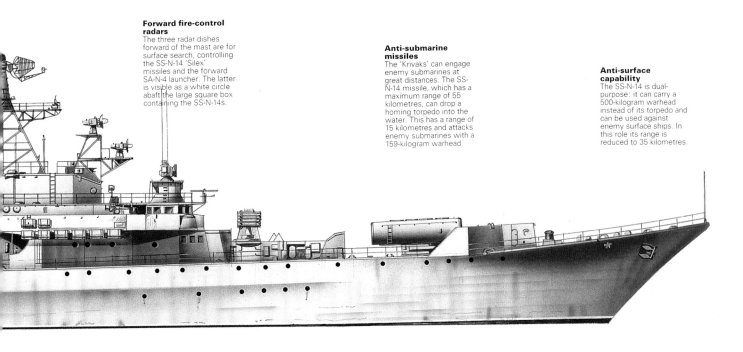

Forward fire-control radars
The three radar dishes forward of the mast are for surface search, controlling the SS-N-14 'Silex' missiles and the forward SA-N-4 launcher. The latter is visible as a white circle abaft the large square box containing the SS-N-14s.

Anti-submarine missiles
The 'Krivaks' can engage enemy submarines at great distances. The SS-N-14 missile, which has a maximum range of 55 kilometres, can drop a homing torpedo into the water. This has a range of 15 kilometres and attacks enemy submarines with a 159-kilogram warhead.

Anti-surface capability
The SS-N-14 is dual-purpose: it can carry a 500-kilogram warhead instead of its torpedo and can be used against enemy surface ships. In this role its range is reduced to 35 kilometres.

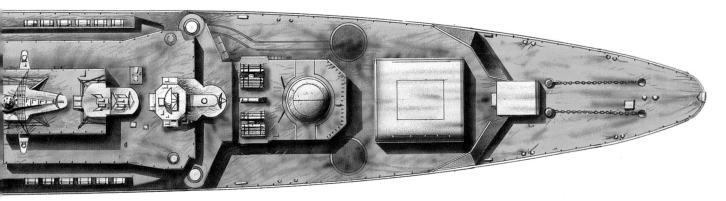

by a new type of fully-automatic water-cooled single-barrel 100-mm gun capable of firing at up to 80 rounds per minute against either air or surface targets. Only 11 'Krivak IIs' were completed compared with 21 examples of the 'Krivak I' type over the construction cycle of the whole class. Surprisingly, units of both sub-classes were completed without their 'Bell' series ECM systems, as a result possibly of the demand for higher-priority shipbuilding programmes.

In 1978 the 'Krivaks' were redesignated *storozhevoy korabl'* (patrol ship). The reason for this may have been either an initial misrating of the type by the Soviets or, more likely, a change in Soviet naval policy during this period. The latter involved a shift of emphasis for surface ASW units, from open-ocean hunter-killer operations against nuclear-powered missile submarines (SSBNs) to a more supportive ASW role in order to secure from interdiction by Western nuclear attack submarines the operating areas of Soviet SSBNs and the transit routes to and from the submarine bases. For this role the 'Krivaks' were ideal,

working in neat conjunction with the medium- and short-ranged shore-based ASW aircraft of the Soviet naval air forces with the aid of information gained from the networks of underwater acoustic arrays that are said to protect the Soviet coastlines. The smaller 'Mirka', 'Petya' and 'Grisha' classes could then perform the more general ASW missions within these 'sanitised' areas and support the 'Krivaks' on lengthy search missions, possibly using a centralised command post afloat or ashore to co-ordinate the sweep.

'Krivak III' launch

The lack of an integral ASW helicopter was found to be very limiting, however, and this resulted in the further revision of the 'Krivak' design to give the 'Krivak III' type, of which the first unit was launched in 1983. This uses the same hull and propulsion plant, but the armament fit has been extensively changed in some areas. The two aft gun turrets have been replaced by a tall and narrow helicopter hangar and flight deck for a single Kamov Ka-25 'Hormone-A' or

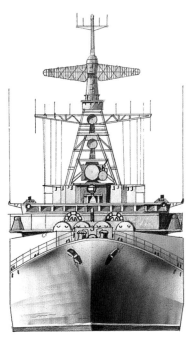

The bow view of a 'Krivak' is dominated by the large masts and radar systems. 'Head Net' air-search radar is fitted above the surface-search radar and the fire-control radars for the ship's missiles.

Ka-27 'Helix-A' ASW helicopter. The variable-depth sonar is retained in a housing now located beneath the flight deck. The minelaying capability is also retained, but the quadruple SS-N-14 'Silex' launcher forward is replaced by a single 100-mm DP gun turret. A 'Kite Screech' fire-control radar replaces the 'Eye Bowl' missile-guidance radars. The forward SA-N-4 'Gecko' SAM launcher bin and associated 'Pop Group' fire-control radar remain, but the corresponding aft systems have been deleted and two 30-mm ADG-6 CIWS fitted aft on each side of the hangar with the associated 'Bass Tilt' fire-control radar forward of the funnel. The remaining ASW armament of two fully automatic RBU 6000 ASW rocket-launchers and the two amidships 533-mm torpedo tubes also appear to have been retained.

Exit the Bosporus

The transit of the first of this class through the Bosporus in the autumn of 1984, from the Soviet Black Sea Fleet via the Suez Canal to the Pacific Fleet, appeared to confirm the importance of the new class for the future of ASW operations in the Far East. In the previous few years the Pacific Fleet had been radically restructured and its capabilities enhanced. However, the 'Krivak IIIs' were not for service with the navy; instead, they were operated by the Border Guards directorate of the KGB.

'Krivak IIIs' continued to be built in small numbers for the KGB right up until there was no more KGB. Six had been delivered by the Crimean Kamysh-Burun yard by 1990, with two more under construction. Production of the 'Krivak II' for the Soviet navy ceased in 1981, however. Clearly, a replacement was long overdue.

The first of the 'Neutrashimy' class of frigates started sea trials in the Baltic in 1990, and two more were under construction when the Soviet Union collapsed. Larger than the 'Krivak', the frigate has a new ASW missile system, and has been designed from the start to operate ASW helicopters. Since frigates are at the less expensive end of the blue-water warship scale, it is possible that the building programme will continue for the Russian navy, when larger, more ambitious designs are scrapped.

Above: Note the relative size of the crew to the giant quadruple launcher for the SS-N-14 'Silex' anti-submarine missiles.

A US Navy P-3 Orion flies over the 'Krivak II' class frigate Zadorny as she heads across the Atlantic bound for Cuba in 1988.

LOS ANGELES

They hunt the deeps of the ocean. Silent, stealthy and deadly. Blind, but with a sense of hearing that would put a bat to shame. One hundred or more men crammed into a 10-metre diameter cigar tube that is loaded with weapons and packed with millions of dollars' worth of electronics. The hunter-killer submarine is like a shark. It is a killing machine, pure and simple.

At every moment of every day, even with the end of the Cold War, a potentially deadly game is being played beneath the North Atlantic between the submariners of the US Navy, the Royal Navy and the former Soviet navy. In the shadowy war of nerves beneath the waves, it is the US Navy's attack submarine force which has for four decades provided the main underwater component of NATO's naval command.

The submarine changed the face of naval combat in two world wars, but it was not until the arrival of nuclear power that the full potential of underwater warfare was unleashed. A modern attack submarine like the American USS *Los Angeles* can remain submerged for months at a time. It can dive deeper and travel faster than at any time in the past.

Submarines are weapon platforms, pure and simple. You can't use a submarine in policing the waterways of the world, and they don't make 'show the flag' port visits. A submarine cruise is the nearest you can get in peacetime to wartime conditions. Submariners have to be masters of their weapons, sensors and tactics: they are the only military men who actually train against the potential enemy.

The 'Los Angeles' class attack submarine USS City of Corpus Christi makes a spectacular sight as it ploughs through the sea. It is very different under water, when the 6,000-ton boat moves like a ghost.

The ultimate hunter-killer

Inside USS *Los Angeles*

'Los Angeles' class boats are faster than almost any other warships in the US Navy. They are also pretty big. A crew of 130 and the living space and supplies to sustain them for months at a time take up a lot of room. When you add a nuclear reactor, a powerplant that could light up a small city, more electronics and computers than you would see on the starship *Enterprise*, and weapons and sensors filling any space left over, then it is clear that there is no room to swing a ship's cat. In fact, there is no room for the cat!

Propulsion
Power is provided by a nuclear reactor aft of the fin. This generates steam to drive a system of turbines delivering 35,000 horsepower. Reduction gears convert the high turbine revolution rate down to 120 revolutions per minute or less, which turns the propeller via the propeller shaft. A large, slow-turning propeller makes less noise than one spinning at high speed.

Control deck
This is the heart of a submarine in action. Red-lit to preserve night vision, it is from here that speed, direction and depth are regulated by helmsmen manning aircraft-like controls. Here, also, the sonarmen have millions of dollars' worth of electronics at their command. The boat fights and weapons are controlled from the attack centre beneath the fin.

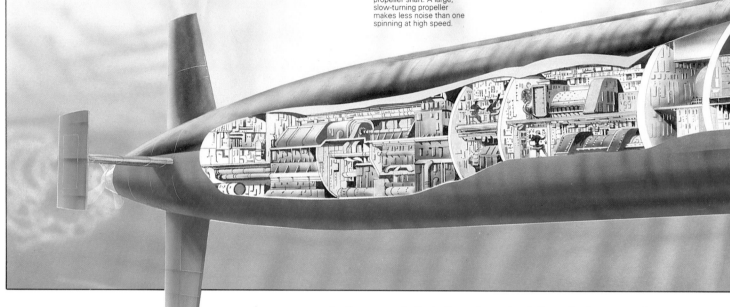

For many years, Soviet ballistic missile submarines were one of the major threats to the West. Early boats were armed with relatively short-range missiles. They had to go far out into the Atlantic to get into firing positions, which made them vulnerable to NATO's anti-submarine forces. In the 1980s, however, things changed. Very long-range missiles meant that the latest Soviet boats hardly needed to leave port to hit any target in the northern hemisphere. And even after leaving port they made life difficult. Soviet missile submarines were built to operate under the Polar ice cap, firing their missiles through the patches of thin ice known as *polynyas*.

Naturally, NATO hunter-killers have to follow their prey wherever they go, and so more and more patrols began to take place in the eerie world beneath the frozen surface of the Arctic. To neutralise the threat posed by the huge 'Typhoon' and 'Delta IV' class submarines, Western hunter-killers also had to be prepared to penetrate deep into heavily-defended Soviet waters, or to follow their targets into their normal operating areas under the Arctic ice.

Advanced designs

Soviet missile boats were only part of the problem. The improvement in submarine performance was marked in the 1970s and 1980s, and new advanced designs appeared with some regularity. By the end of the 1980s, Soviet attack boats often had better performance than Western designs, and NATO boats would have had to run a gauntlet to reach the missile boat targets.

If it had ever come to a shooting war, it would have been submarines like USS *Los Angeles* (SSN 688) that would have had to penetrate the heavily-defended Soviet naval sanc-

*The sonar team aboard the **USS** La Jolla is listening to the output from the boat's sensitive and sophisticated sensors. It may look like a Hollywood starship, but the attack centre of a modern submarine is a very serious place.*

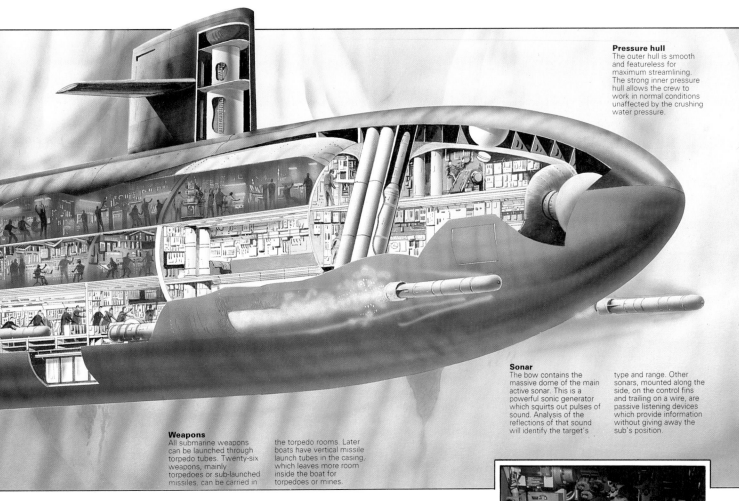

Pressure hull
The outer hull is smooth and featureless for maximum streamlining. The strong inner pressure hull allows the crew to work in normal conditions unaffected by the crushing water pressure.

Sonar
The bow contains the massive dome of the main active sonar. This is a powerful sonic generator which squirts out pulses of sound. Analysis of the reflections of that sound will identify the target's type and range. Other sonars, mounted along the side, on the control fins and trailing on a wire, are passive listening devices which provide information without giving away the sub's position.

Weapons
All submarine weapons can be launched through torpedo tubes. Twenty-six weapons, mainly torpedoes or sub-launched missiles, can be carried in the torpedo rooms. Later boats have vertical missile launch tubes in the casing, which leaves more room inside the boat for torpedoes or mines.

tuaries in order to detect and destroy the Soviet missile submarine fleet. At the same time, other members of the class would have been duelling with Soviet hunter-killers and cruise missile boats, preventing them from breaking out into the Atlantic and staving off attacks on the US Navy's carrier battle groups and NATO's merchant lifeline.

Although the Cold War has ended, the Russians still control a vast and powerful navy. As long as the threat of instability remains in the former Soviet Union and that navy remains in existence, NATO and the West must be ready to counter its activities. That readiness depends above all else on the US Navy's 'Los Angeles' class attack boats.

With more than 50 boats in service, building and on order, and with more to come, the 'Los Angeles' or '688' class comprises the largest number of nuclear warships ever built to one design. They are faster than other Western submarines, but slower than the latest opposition. When the first of the superfast Soviet 'Alfa' class ventured out into the Atlantic, it was able to outrun a shadowing 'Los Angeles' boat with ease. But pure speed is not that important.

The problem with travelling fast is that you can't hear anything over the noise of your own passage through the water. The only sense of use to a submariner is the sense of hearing. Sound tells you where you are and tells you where the enemy is. It is vital to control the sounds you generate: in battle, a noisy submarine is a dead submarine.

And this is where American boats

Right: The duty watch in the control room of USS La Jolla monitors the submarine's depth, bow angle and speed as it patrols beneath the Pacific.

Torpedoes are large objects and loading them aboard a submarine can be a complex process.

'Los Angeles' class submarines are often attached to carrier battle groups in high-threat environments, where they serve as highly capable additions to the group's ASW defences.

longer ranges they have tube-launched missiles, which break the surface and fly to deliver their payload of a torpedo or nuclear warhead onto an unsuspecting submarine many miles away. Different missiles can be carried for long-range ship- and land-attack.

Variety of weapons

In spite of their size, modern subs don't carry too many weapons. It is not surprising when you consider that a heavyweight torpedo weighs in at one and a half tons and is almost six metres long! Although it is one of the largest attack submarines around, the 'Los Angeles' only carries 26 torpedo tube-launched weapons. As a result, the weapons that are carried have to be able to deal with a variety of tactical situations. These are usually a mix of Mk 48 ADCAP wire-guided torpedoes, Sub-Harpoon anti-ship missiles, SUBROC anti-submarine missiles (which were to be replaced by the new Sea Lance missile in the 1990s before Congress cut off funding), Tomahawk nuclear- or conventionally-armed cruise missiles, or mines. To alleviate the storage problem, later boats have been fitted with vertical launch tubes in the outer casing for up to 15 Tomahawks, allowing more anti-submarine weaponry to be carried in the torpedo room.

Although there are faster submarines around and future craft will be even quieter, the '688' boats have a combination of speed, stealth and outright fighting ability that is hard to beat. The first entered service in 1976, and the last will probably be commissioned in the late 1990s. With the cutback in defence spending and the cancellation of the US Navy's advanced 'Sea Wolf' submarine programme, 'Los Angeles' class boats will remain at the forefront of the West's defences for the foreseeable future, serving well into the 21st century.

score: at normal operational speeds a '688' boat is so silent that it has been described only half facetiously as 'like a hole in the water'. Stealth might be a new buzzword to the air forces of the world, but it has been a prime requirement for underwater warfare for decades.

With such a quiet platform the crew can concentrate on trying to identify the enemy. All submarines have passive sonar suites, which consist of sensitive hydrophones at the bow, along the sides and towed astern of the boat. They are listening for the give-away noises generated by a target's reactor cooling systems, its engines and the churning of displaced water caused by the propeller. Submariners use a technique called sprint and drift. They move at high speed for a short period and then cut the power right back to listen for a threat. If all is clear they will make another sprint and repeat the process.

Sophisticated electronics

With the most sophisticated electronics in existence making sense of the data coming in from the boat's sonars, the skipper of a 'Los Angeles' knows more about the mysterious underwater world around him than any submarine commander in history. This makes the '688' boats superb at detecting and tracking other sub-

marines without the opposition's knowledge.

The dome in the front of a 'Los Angeles' class submarine is a powerful sound generator, part of the boat's active sonar system. It squirts out bursts of sound, which echo back from anything solid in their path. Analysis of the reflected sound allows the submarine to classify the target and provides enough information to identify it. It can even differentiate between different types of submarine. However, although they can give detailed information, submarines rarely use their active sonar because squirting out powerful beams of sound is not a good idea if you are trying to be stealthy.

Knowing about something and having the capability to deal with it are two different things. Submarines have a choice of weapons. Against submarines at relatively close range they can use torpedoes, which are also effective against surface ships. For

The USS Birmingham emerges from the ocean in a dramatic welter of spray during an emergency surfacing exercise. The normal surfacing procedure is much less spectacular.

MAD Systems

Magnetic Anomaly Detectors (MADs) are widely used by maritime aircraft and helicopters to pinpoint enemy submarines. The magnetic field of the Earth is distorted by the presence of any large mass of metal. Ships and submarines exert a pronounced local effect on the magnetic field as they become magnetised during construction. Although it has long been common practice to demagnetise warship's hulls from time to time – thus reducing the threat from mines detonated by disturbances in the magnetic field – there is no way of completely eliminating the problem.

The magnetic properties of submarines were understood before World War I, and various systems were built to detect a submerged submarine in this way. However, most of them were static shore-based systems – one was incorporated into the defences of the Royal Navy anchorage at Scapa Flow. The range at which a submarine can be located by a MAD is very limited, even today. Modern

The RAF's Nimrod maritime patrol aircraft also have a MAD system projecting from the tail. The instruments need to be isolated from other equipment aboard the aircraft that affects the magnetic field.

systems can pick up a submarine at shallow depth at a range of no greater than 2000 metres, so as a primary means of detection MADs are of limited value. Their true value is to confirm the exact location of a submarine that is being monitored by passive sonar, so the target can be attacked before it even knows it has been located.

Shadowing technique

Maritime patrol aircraft, such as the Nimrod or P-3 Orion, can detect and track submarines without the target knowing it is being shadowed. By dropping passive sonobuoys into the sea the aircraft can discover the bear-

A US Navy Lockheed P-3 Orion flies over a submarine that has its masts raised above the surface. The P-3's extended tail houses the MAD instruments.

ing of the submarine. But it does not reveal the range – the submarine could be right next to the buoy or several thousand metres away. Sound travels in mysterious ways under the sea and the volume of underwater noise is seldom a reliable indication of the proximity of the source. Aircraft can drop patterns of buoys and use a cross-bearing to pinpoint the submarine. But by flying low over the water they can use a MAD to do the same. This is especially valuable for

naval helicopters with limited supplies of sonobuoys.

A submarine will be detected by a MAD at up to about 2000 metres unless it has gone deep, in which case the device will only work at up to approximately 1000 metres. There is no defence against it. It is just as reliable against a submarine running silent – creeping at less than five knots with all non-essential machinery shut down – as it is against a nuclear-powered boat surging through the ocean at flank speed.

Soviet countermeasures

Only the Soviet navy introduced submarines that were largely immune to Magnetic Anomaly Detectors. The innovative 'Alfa' class nuclear-powered attack submarines were constructed with hulls of titanium alloy, which is notoriously difficult to weld

The masts of a Soviet 'Victor I' class submarine. These SSNs produce a significant magnetic anomaly.

but immensely strong and has little magnetic influence. With their small but incredibly tough pressure hulls,

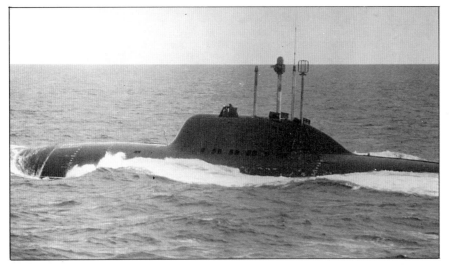

Above: One of the Soviet navy's titanium-hulled 'Alfa' class nuclear submarines that can dive too deep to be detected and exert less influence on the magnetic field.

the 'Alfas' are capable of diving to 700 metres in safety. A submarine lurking in such great depths cannot be detected by an airborne MAD. By comparison, the new British 'Trafalgar' class are credited with diving depths of 300 metres; the American 'Los Angeles' class boats can go to 450 metres. While allowing for the fact that the Royal Navy does not like to reveal the full capabilities of its ships, even to the most respected reference books, it is accepted that the 'Alfas' can out-dive any rival submarine.

The only good news for NATO anti-submarine forces is that the 'Alfas' are built to a design now over 25 years old, and they are noisy boats, lacking the anechoic coating in which more recent types have been clad. Yet, there is a successor type, the 'Sierra' class, also credited with an impressive diving depth thanks to a titanium-strengthened hull.

Below: A US Navy S-2 Tracker with its retractable MAD boom extended. The MAD is used as final confirmation before an attack, since a low-flying aircraft may be detected by a submarine.

NAVAL MINES:
Death from Below

Above: A simulated mine explosion alongside a US Navy frigate shock tests the vessel's structure. If directly underneath, the detonation would have broken the frigate's back.

Above: The Versatile Exercise Mine System is used by the Royal Navy to simulate many different types of modern mine.

Below: Modern mines are small and easy to handle, and can be laid from almost any platform without the need for special handling equipment.

Naval mine technology has evolved at a phenomenal rate in the last few decades, a product of the computer and electronic revolution that has so changed the face of warfare. The modern mine bears little resemblance to the horned monsters that were so feared by sailors in earlier days and that are familiar from Hollywood war movies, although they are still in use. Nevertheless, the purpose of mines remains unchanged: to deny an enemy free use of the seas.

Mines are effective for a number of reasons. The seas cover 70 per cent of the world's surface, and mining every possible sea route would be impossibly expensive. However, there are certain areas where you know that ships are going to be. Every ship has to come into harbour, and in any country, no matter how big, there are only a limited number of ports. Mining the approaches to those ports effectively strangles trade. There are also natural choke points. For example, Soviet access to the open Atlantic is limited by Arctic ice, the Greenland/Iceland/UK Gap, the Skagerrak, the Bosporus and the Straits of Gibraltar, all of which can be mined with greater or lesser difficulty.

Similarly, a large part of the world's oil comes through the Straits of Hormuz at the mouth of the Persian Gulf. Much of the Far East's trade passes through the Straits of Malacca, and without the Panama and Suez canals the world's shipping would be faced with long and costly passages around the Cape of Good Hope and Cape Horn. Mines at any of these locations could have a serious effect out of all proportion to the cost of the weapons involved.

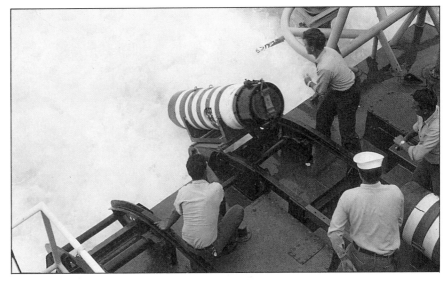

Minefields are easy to lay at very short notice. Mines, or more precisely mine detonation systems, generally come in four different types. Contact, magnetic, acoustic and pressure actuated mines are very different, and minesweeping techniques capable of handling one type of mine will have little or no effect on the others.

Contact mines should be anchored so that they float just below the surface of the water, and as their name implies they require physical contact between ship and mine in order to be detonated. The other types are described as influence mines, designed to lie on the sea floor. Exploding at a distance from the target ship, they are designed to attack from beneath the keel, the shock wave of their detonation breaking the ship's

back. Magnetic mines are activated by changes in the Earth's magnetic field caused by a large piece of metal – a ship – passing overhead. Acoustic mines are detonated by sounds, actuated by the combination of noise from the engines, the propellers and the flow of water over the hull of a ship passing overhead.

The size of mines vary. The old-style horned contact mines are best laid from ships, but other types which usually lie on the seabed are smaller, and can be laid by aircraft, helicopters, warships, civilian vessels or submarines.

Mines are amongst the most effective yet underrated weapons in modern warfare. They are just as terrifying to modern sailors as they were to their fathers and grandfathers.

Engagement sequence

1 Contact mines

Contact mines are the oldest and simplest form of underwater explosive device, which as the name implies are detonated by ships making physical contact with the mine. Contact mines were used in the American Civil War, when they were known as torpedoes. By the time World War I came around they had achieved their present form. Moored to the sea bed, they are generally spherical or ovoid, with a number of horns which contain the electrical detonators. The horns are simple batteries. When they are broken open, either seawater completes the circuit, or acid is released from a container. In each case, a current flows and an electrical firing charge is generated. Explosive fillings vary, but are usually between 100 kg and 250 kg of high-grade explosive such as TNT, Torpex or Amatol.

2 Pressure/magnetic mines

Pressure and magnetic mines are what are known as influence weapons, being able to detect and destroy a target without the need for physical contact. Both are sensitive to changes in their immediate environments caused by the passage of a ship overhead. Both types of mine can be programmed to deal with selected targets, ignoring destroyers and frigates in order to take out the huge bulk of an aircraft carrier. A ship's hull passing through the water has the effect of decreasing the water pressure immediately beneath it, and a pressure mine can sense that pressure difference. Magnetic mines are activated by a change in the Earth's magnetic field caused by the proximity of a large metallic conductor nearby, and whatever else a ship is, it is almost certainly going to be a large chunk of metal.

3 Acoustic mines

Acoustic mines are activated by noise. Even though modern warships have been silenced to a considerable extent, no vessel can be totally quiet. Anything moving through the ocean will generate sound, as much from its hull passing through the water as from its propellers and engines. This noise is picked up by sensitive hydrophones on the mine. The original acoustic mines developed before World War II were not very discriminating, but they could be fitted with ship counters, only blowing up when a specified number of ships had passed by. Modern mines have this facility, but they also have many other tricks. Every vessel has its own unique sound signature, and acoustic mines can be programmed to respond only to a specific type of ship or submarine.

4 Rising mines

Rising mines have been developed to extend mine capability. Most mines are limited to fairly shallow water. Moored mines can be found in 2,000 feet of water, although shallower water is more usual. The mine itself will float between 30 and 300 feet down. Ground mines are usually laid no more than 200 feet down, because for maximum effect they should detonate within a hundred feet or so of a target ship. Rising mines are not so limited: they can be laid in very deep water. Once a rising mine detects a target, it releases a projectile which is often a torpedo that rises up to strike the target. Using a torpedo also means that the target does not have to pass overhead, but can be engaged at a distance. The effect of deep-water rising mines is to dramatically increase the area of the world's oceans which it is now possible to mine.

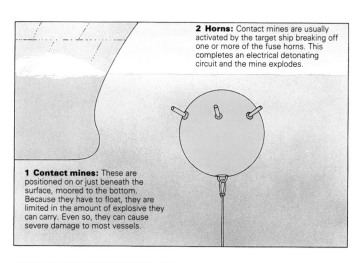

2 Horns: Contact mines are usually activated by the target ship breaking off one or more of the fuse horns. This completes an electrical detonating circuit and the mine explodes.

1 Contact mines: These are positioned on or just beneath the surface, moored to the bottom. Because they have to float, they are limited in the amount of explosive they can carry. Even so, they can cause severe damage to most vessels.

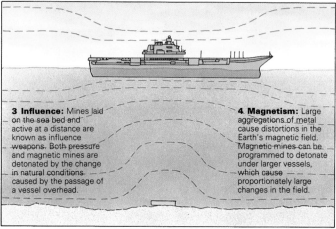

3 Influence: Mines laid on the sea bed and active at a distance are known as influence weapons. Both pressure and magnetic mines are detonated by the change in natural conditions caused by the passage of a vessel overhead.

4 Magnetism: Large aggregations of metal cause distortions in the Earth's magnetic field. Magnetic mines can be programmed to detonate under larger vessels, which cause proportionately large changes in the field.

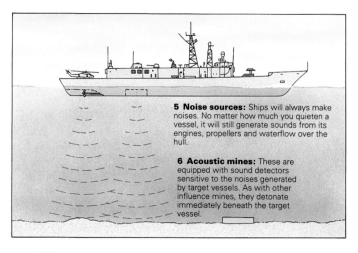

5 Noise sources: Ships will always make noises. No matter how much you quieten a vessel, it will still generate sounds from its engines, propellers and waterflow over the hull.

6 Acoustic mines: These are equipped with sound detectors sensitive to the noises generated by target vessels. As with other influence mines, they detonate immediately beneath the target vessel.

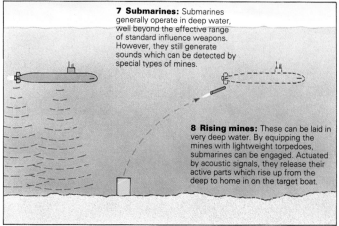

7 Submarines: Submarines generally operate in deep water, well beyond the effective range of standard influence weapons. However, they still generate sounds which can be detected by special types of mines.

8 Rising mines: These can be laid in very deep water. By equipping the mines with lightweight torpedoes, submarines can be engaged. Actuated by acoustic signals, they release their active parts which rise up from the deep to home in on the target boat.

OTO-MELARA 76-mm

The OTO-Melara 76-mm Compact gun is one of the most widely used naval guns in the Western world. It is dual purpose, i.e. it can engage both air and surface targets.

The pre-eminence of the aircraft carrier in the immediate post-war years, together with the rise of the guided missile, seemed to signify the end of the naval gun as a major weapon system. Indeed, the Royal Navy went so far as to commission a major class of warship, the 'Type 22' frigate, without a gun. However, events in the 1980s and 1990s have shown the error of such thinking, and the naval gun has undergone a revival.

Although modern fighting vessels are rarely likely to use their guns in combat with another warship, the Falklands war and the multinational intervention in the Lebanon in the early 1980s showed that there was still an extremely important place for naval artillery, particularly when providing naval gunfire support for troops ashore. The Gulf War of 1991 only confirmed the trend, with the allies being able to call on the awesome power of the US Navy's revitalised battleships.

Guns at sea

Big guns are very expensive, however, and by far the largest number of such weapons at sea are of medium calibre – between 76 mm and 125 mm. France, Russia, Sweden and the USA are major players in the export of medium-calibre naval artillery, but it is the Italian 76-mm Compact gun, manufactured by OTO-Melara, which is the outstanding success in the field.

Developed from the OTO MMI single gun mount, the OTO-Melara 76-mm Compact gun is the most widely used naval weapon of its type in the world. It entered service in 1969 as a dual-purpose anti-air/anti-surface system intended to provide secondary armament to major warships or to act as primary armament of smaller vessels down to the size of fast attack craft and hydrofoils.

The mounting consists of two parts, the shank and the turret assembly. The shank is the below-deck part of the installation and contains a rotating platform or drum which can hold up

US Navy frigates of the 'Oliver H. Perry' class have a single OTO-Melara 76-mm gun mounted on top of the superstructure.

to 80 rounds, together with a hoist ammunition feed system which delivers rounds to the turret above. The turret contains the gun and its elevation training mechanisms. The equipment is covered by a watertight and NBC-proof glass-reinforced plastic shield.

The gun barrel is 62 calibres long, and is fitted with a muzzle brake and a fume extractor. The 76-mm Compact fires a 6.3-kilogram projectile with a muzzle velocity of nearly 1000 metres per second

Fully automatic gun

The 76-mm Compact system is fully automatic. The only manpower required is below deck around the magazine, where ammunition handlers are needed to keep the ammunition drum loaded. The rate of fire is adjustable, from a minimum of 10 rounds per minute to a maximum of 85 rounds per minute.

The gun is aimed and fired through the ship's fire-control system, and rate of fire is selected on a console in the operations room or combat control centre. If necessary, however, the 76-mm Compact can be fitted with a stabilised line-of-sight aiming system for emergency manual control.

The 76-mm Compact is one of the most successful naval weapons to have been developed since World War II. Currently in service with nearly 40 navies, the light weight of the system enables it to be installed on

An Egyptian fast attack craft of the 'Ramadan' class, built by Vosper Thorneycroft in Britain. It is armed with a 76-mm Compact gun forward and twin 40-mm anti-aircraft guns aft.

vessels displacing as little as 60 tonnes. The 76-mm system has seen extensive combat use with the Israeli navy. It is being manufactured under licence by the United States, where it forms the gun armament on some 60 'Oliver Hazard Perry' class frigates, as well as on the six 'Pegasus' class hydrofoils. Other manufacturing countries include Australia, Japan and Spain.

The 76-mm system was further de-

veloped as the 76/62 Super Rapid gun mounting. This incorporates a number of improvements, most notable is the higher maximum sustained rate of fire of 120 rounds per minute, which together with new pre-fragmented high-explosive ammunition gives the system better capability against sea-skimming missiles. The round is fitted with a proximity fuse, and it can also be used as an anti-personnel round in shore bombardment.

Specification

OTO-Melara 76-mm Compact gun

Calibre: 76 mm
System weight: 7.35 tonnes
Projectile weight: 6.3 kg
Muzzle velocity: 925 m/sec
Elevation: −15°/+85°
Maximum rate of fire: 85 to 100 rounds per minute, depending upon model
Effective range: 8 km in surface action and 5000 m in anti-air role

Istiqlal was the only Lürssen-built FPB 57 type fast attack craft of the Kuwaiti navy to survive the Gulf War, escaping to Bahrain while the rest were captured by Iraq. Note the 76-mm Compact gun in the forward turret.

PASSIVE SONAR

A US Navy sonar operator compares a recent contact to a manual of recorded sonar signatures. Modern systems can determine the exact class of warship they have detected.

Modern underwater detection systems have progressed a long way from the simple hydrophones of the 1920s and the ASDIC used to hunt German U-boats during World War II. Employing the same principle of using sound to penetrate the otherwise impenetrable depths of the oceans, modern electronics give sonar a range and discrimination undreamed of in earlier years.

How well a sonar performs is effectively the limiting factor in almost all ASW systems, apart from non-acoustic detectors such as MAD – Magnetic Anomaly Detection – gear. Sonar is divided into two major types.

Passive sonar is simply listening gear, although the equipment currently in use is far from simple. It incorporates extremely sensitive detectors that are used to detect the smallest sound in the oceans around the receiver.

Active sonar requires a powerful sound generator. It sends out an acoustic energy pulse or 'ping' into the surrounding water and the receiver waits for an echo to return from the target.

There is a general rule of thumb that the lower the frequency of the sound waves used then the greater the range a target can be detected. There are rumours that the low-frequency sounds generated by the super-fast Soviet 'Alfa' class submarines were detected by sensors at AUTEC, the US Navy's underwater test range off the Bahamas, which is a distance of over 5,000 miles!

The trouble with active sonars is that, by definition, they are very noisy. After all, they squirt a powerful beam of sound into the water. Your target is certainly going to hear that long before you can hear the faint echo coming off his hull.

It is for this reason that submarines rarely use the active sonar with which they are equipped, except possibly in the last stages of an attack. Submarines are the business of clandestine conflict: they are the original 'stealth' fighting machines. Sub-

HMS Spey's sonar enabled her to sink two U-boats in 1944. World War II sonar had a range of a few thousand yards; modern passive systems can detect targets at 100 miles in good conditions.

marines are most effective at making surprise attacks, and announcing your intentions by yelling at the top of your voice is not a very good way of retaining the element of surprise.

Although the sea is filled with sound, passive sonar can generally be used to distinguish between natural and man-made noises. Naturally-generated sounds range from the subsonic rumblings caused by displaced magma deep beneath the sea bed to the supersonic clicks and squeaks of dolphin echo-location. Such sounds are rarely regular in pitch or intensity.

Radiated noise

Man-made sounds are distinctive, however. Any ship will send out a signal into the surrounding water mass in the form of radiated noise. This can be picked up by a hunting submarine using its sensitive hydrophone arrays.

The sound signature of any vessel has several major components. The engines produce repetitive pulses of noise, and propellers are a major noise source: cavitation, the bubbles a propeller produces as it whips through the water, is easily detectable at long ranges. Finally, the noise of water flowing over the hull also adds to the noise signature.

For successful detection, the incoming signal must be sufficient in amplitude to be distinguishable against the background noise of the sea. It must be separated also from extraneous noises made by the listening platform itself, which after all is another man-made object.

Modern passive sonar systems use very sensitive sensors allied to extensively computerised signal-processing units. These digitalise sound and subject it to the kind of filtering process that would be beyond the dreams of the world's most expensively-equipped recording studio. The systems can isolate extremely faint signals even through high ambient noise, and under favourable conditions they can do so at ranges in excess of 150 kilometres.

In order to maximise sensitivity, sonars are often isolated from the noises generated by their own platform by means such as extensive and expensive sound deadening. However, the most effective measure is the towed array, which involves towing the sonar array behind the boat on a long line, well clear of sound contamination.

The trouble with passive sonars is that a passive sonar user is very much

The deep-diving Soviet 'Alfa' class attack submarines may have been incredibly fast but they made a great deal of noise. The US Navy's SUBROC nuclear missiles were developed so that American SSNs could destroy Soviet boats detected at long range with passive sonar.

at the mercy of the capriciousness of nature. The sea is not a uniform, static place. It has many currents: some local, some seeping for thousands of miles. These currents create layers of water of differing density, temperature and salt content.

Under certain conditions and from certain angles the boundary between two layers will reflect sound. A submarine beneath such a layer is unlikely to be detected by a ship-mounted or helicopter-mounted sonar. Submarines are very aware of such layers, and a switched-on skipper will use these features to evade detection.

The sonar team of a Royal Navy submarine in action. In 1982 HMS Conqueror shadowed the Belgrano and its escorts using passive sonar. Fear that the cruiser was heading into the shallow waters over the Burdwood bank helped spur the decision to attack.

PENGUIN
Anti-ship Missile

The original development of the Penguin took place in the 1960s, and was a collaborative effort between the Royal Norwegian Navy, the Norwegian defence establishment and the Norwegian defence industry. The requirement called for an anti-invasion weapon system, compatible with small- and medium-sized vessels operating in coastal waters.

It was the first Western anti-ship missile, and in 1972 it went into service on 'Storm' and 'Snogg' class fast attack craft. Ship-launched Penguin is carried in a simple container/launch box that weighs about 650 kilograms complete with missile. The box is located on a prepared deck mount that has an umbilical connection to the fire-control system. Once the missile is connected, it is ready to launch.

Penguin 2 is an improvement of the original design, with a better seeker head and electronics; it retains all of the original missile's characteristics, however. These include an engagement range comparable with the maximum effective radar range for fast patrol boats, and a completely passive fire-and-forget operation.

Missiles which are active or semi-active homing can have their radar signals interfered with by the enemy, but there is little the foe can do with completely passive systems. Launched on a target bearing, Penguin uses inertial navigation until the terminal phase of an attack, when it switches to infra-red homing and is guided by the target's own heat output.

Apart from shipborne use, there are also Penguin variants for coastal defence and for air-launch from helicopters. These are similar to the ship-launched version. The Penguin 3, however, is designed for aircraft use. It has smaller, folding wings, and because it has no booster it must be launched from a platform travelling at 200 knots (370 km/h) to be effective.

The US Navy is fielding a folding-winged variant of the Penguin 2 on its SH-60 Sea Hawks to give an anti-fast patrol boat capability. Each helicopter can carry two missiles.

All versions of the Penguin are equipped with a modified Bullpup warhead, a 120-kilogram semi-armour-piercing round with an impact fuse. This is enough to destroy fast patrol craft, and can also damage larger vessels.

Below: A Penguin Mk 2 missile flies low over the sea's surface after being launched from an 'Oslo' class frigate of the Royal Norwegian Navy (main picture). Penguin is a fire-and-forget missile, navigating inertially to the target area before homing in on the heat of the target vessel.

Specification

Penguin Mk 2

Type: ship-, shore- or helicopter-launched anti-ship missile
Dimensions: length 2.96 m; diameter 28 cm; wing span 1.4 m
Range: 30-40 km
Speed: Mach 0.8
Guidance: inertial with terminal infra-red homing
Warhead: modified Bullpup, 120-kg high-explosive semi-armour piercing

Engagement sequence

1 Platform

Initial target surveillance, detection and designation are carried out by the launch platform's systems. After detection by radar, the fire-control system will calculate the speed and direction of the target, it will work out where the target will be by the time the missile has reached that far, and will predict the point of impact. The missile's inertial guidance system is slaved to the vessel's fire-control system until the moment of launch. On firing, the missile will head off in the general direction of the target. Target designation to the Penguin 3 air-launched missile can be achieved by placing a marker over the target in the aircraft's radar display, or on the head-up display, and pressing a button to transmit the data to the missile.

2 Fire-and-forget

After launch, Penguin uses inertial guidance to perform the cruise phase of the flight, with infra-red homing in the terminal phase. This means that the missile is independent of the launch platform, which is therefore free to take immediate evasive action. The vessel is also free to immediately engage another target. The same is true of air-launched Penguin platforms. The fire-and-forget facility is particularly valuable to helicopters, since it reduces exposure time to a minimum, reducing the helicopter's vulnerability. Penguin-equipped SH-60 LAMPS III helicopters of the US Navy can thus make autonomous attacks without risk, or they can take part in co-ordinated attacks with ships firing the much larger Harpoon missile from over the horizon.

3 Cruise

After launch or release, Penguin missiles follow a pre-programmed flight path towards the predicted impact point, using their inertial navigation systems and an altimeter for reference. Penguin can fly directly to the predicted target area, or it can be programmed to fly to a waypoint from where it can be instructed to change course, heading in on the target from a direction up to 90 degrees away from the true direction of the launch platform. Penguin can also be programmed to change its altitude at the waypoint, to adopt a sea-skimming flight profile for the terminal phase. Once in the target area, the missile switches from inertial navigation to terminal homing.

4 Terminal phase

Once the Penguin missile reaches the predicted target area, it switches on its infra-red homing system. This is a very sensitive detector which can detect the heat released by a target. The infra-red seeker is in search mode and scans a sector ahead of the flight path. Penguin can follow one of a number of search patterns. Once the target is detected, the seeker head locks on, generating signals which are used to steer the missile in for the kill. Penguin's 120-kilogram semi-armour-piercing warhead was originally designed in the 1950s, when most ships still had some armour protection. It is more than enough to destroy a coastal craft, and can cause significant damage to ships up to destroyer size.

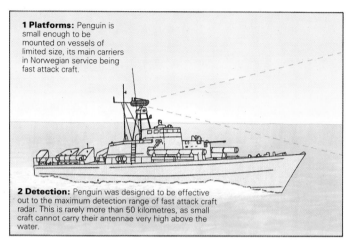

1 Platforms: Penguin is small enough to be mounted on vessels of limited size, its main carriers in Norwegian service being fast attack craft.

2 Detection: Penguin was designed to be effective out to the maximum detection range of fast attack craft radar. This is rarely more than 50 kilometres, as small craft cannot carry their antennae very high above the water.

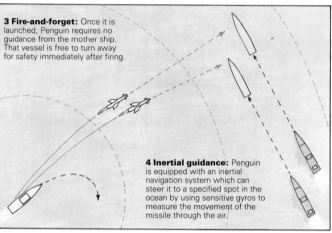

3 Fire-and-forget: Once it is launched, Penguin requires no guidance from the mother ship. That vessel is free to turn away for safety immediately after firing.

4 Inertial guidance: Penguin is equipped with an inertial navigation system which can steer it to a specified spot in the ocean by using sensitive gyros to measure the movement of the missile through the air.

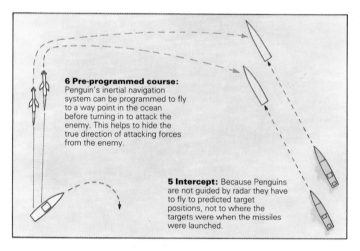

6 Pre-programmed course: Penguin's inertial navigation system can be programmed to fly to a way point in the ocean before turning in to attack the enemy. This helps to hide the true direction of attacking forces from the enemy.

5 Intercept: Because Penguins are not guided by radar they have to fly to predicted target positions, not to where the targets were when the missiles were launched.

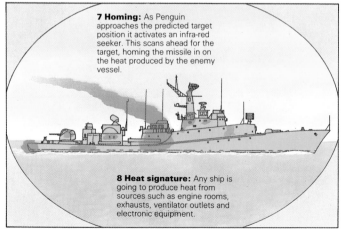

7 Homing: As Penguin approaches the predicted target position it activates an infra-red seeker. This scans ahead for the target, homing the missile in on the heat produced by the enemy vessel.

8 Heat signature: Any ship is going to produce heat from sources such as engine rooms, exhausts, ventilator outlets and electronic equipment.

PHALANX
CLOSE-IN DESTROYER

Phalanx is a ship's last line of defence against an incoming missile. A few seconds before impact, Phalanx's radar guides the six-barrelled 20-mm cannon onto the target. Firing 50 rounds per second, Phalanx destroys the missile and saves the ship.

Fighting ships are vulnerable to attack from the air. During World War II, the solution was to crowd as many light and heavy calibre guns onto the decks of a warship as she could carry.

The rapid development of the surface-to-air missile (SAM) in the years following World War II revolutionised the defences of ships at sea, doing away with the need for dozens of guns and the tens or even hundreds of crewmen manning them. However, the threat evolved just as rapidly, and manned aircraft gave way to anti-ship missiles, their high speeds and small sizes presenting shipborne anti-aircraft weapons with much more difficult targets. The problems were heightened when missiles began to be programmed to make sea-skimming attacks, flying in a few metres above the water.

No missile defence is impregnable, and there are always attackers that are going to get through. The US Navy decided that what it needed was a last-ditch defence, that could lay a wall of shells into the path of an incoming missile in the last few hundred metres before hitting the target. Somehow the dozens of cannon carried by battleships in World War II would have to be replaced.

The General Dynamics Phalanx CIWS (Close-In Weapons System) is a fast-firing, radar-directed cannon that does the job. Based on the awesome firepower of the six-barrelled Vulcan cannon, developed for supersonic fighter aircraft, the Phalanx system can spit out a stream of extremely dense DU (depleted uranium) shells at a rate of more than 50 rounds per second, laying bursts into the path of an attacking missile. The incoming missile and outgoing shells are both tracked by the fire control radar, which adjusts the aim of the gun to ensure that the target is destroyed as far away from the ship as possible.

Specification
Phalanx Mk 15/16 CIWS

Calibre: 20 mm
Number of barrels: six
Elevation: −25° to +80°
Effective range: between 500 and 1500 m
Rate of fire: 3,000 rounds per minute
Used by: Australia, Greece, Israel, Japan, Pakistan, Portugal, Saudi Arabia, Taiwan, UK, USA

The large white dome makes the Phalanx an easy weapon to identify on a warship. It houses the radar and fire control system.

Engagement sequence

1 'Leakers'

Modern fighting ships are extremely expensive investments, so it makes sense to protect them as well as possible. US Navy supercarriers are the most valuable targets afloat, so they are also the most heavily defended against air attack. They lie at the centre of a layered defence, covering a circle hundreds of kilometres in diameter. The outer ring consists of F-14 Tomcat fighters patrolling some 350 kilometres from the carrier. They are armed with long-range AIM-54 Phoenix missiles, which extend their range out a further 150 kilometres. The central band of defence out to 150 kilometres is covered by SAMs from the carrier's escorts. Should any hostile aircraft or missile successfully run that deadly gauntlet, it must be intercepted before it hits the target. These 'leakers' are dealt with by last-ditch defences known as CIWS, or Close-In Weapons Systems.

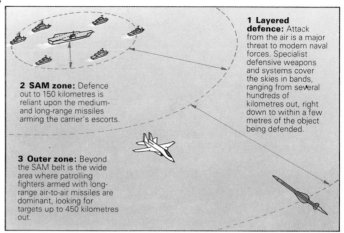

1 Layered defence: Attack from the air is a major threat to modern naval forces. Specialist defensive weapons and systems cover the skies in bands, ranging from several hundreds of kilometres out, right down to within a few metres of the object being defended.

2 SAM zone: Defence out to 150 kilometres is reliant upon the medium- and long-range missiles arming the carrier's escorts.

3 Outer zone: Beyond the SAM belt is the wide area where patrolling fighters armed with long-range air-to-air missiles are dominant, looking for targets up to 450 kilometres out.

2 Detection

Although the layered defence is very capable, it is not an impenetrable shield. The fighters, for instance, are responsible for watching over more than three-quarters of a million square kilometres of ocean, and even with Hawkeye flying radar stations it is possible for isolated attackers to slip through. The escorts in the inner layer depend on shipboard radars to detect targets, which are limited to detecting targets above the horizon. If an incoming missile takes a sea-skimming flight path, it might get to within 20 kilometres – one minute of flying time – before being detected. Sea skimmers are notoriously hard to engage with SAMs, most of which have an engagement envelope which starts at about 20 metres. Target information is fed to the ship's CIWS, just in case the SAMs fail to make an intercept.

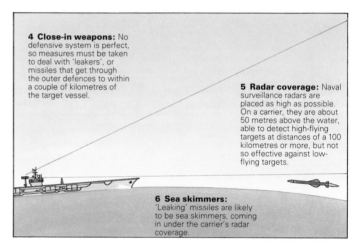

4 Close-in weapons: No defensive system is perfect, so measures must be taken to deal with 'leakers', or missiles that get through the outer defences to within a couple of kilometres of the target vessel.

5 Radar coverage: Naval surveillance radars are placed as high as possible. On a carrier, they are about 50 metres above the water, able to detect high-flying targets at distances of a 100 kilometres or more, but not so effective against low-flying targets.

6 Sea skimmers: 'Leaking' missiles are likely to be sea skimmers, coming in under the carrier's radar coverage.

3 Engagement

Phalanx is a totally independent weapons system. It is equipped with a high-speed digital computer, which can use the system's pulse-Doppler radar to perform search, acquisition and tracking functions. Its primary task is to destroy 'leakers' far enough out to prevent the damaged missiles or wreckage going ballistic and crashing into the target ship. At the same time, Phalanx should not waste ammunition while other, longer ranged systems are still effective. Phalanx is therefore designed to be effective while the incoming missile is less than five seconds from impact, at ranges from 1500 metres down to 500 metres. Maximum lethality occurs at 500 metres. The whole Phalanx assembly moves as a unit, and is capable of destroying terminally diving targets as well as sea skimmers.

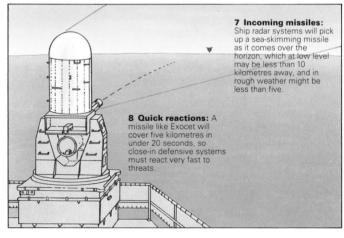

7 Incoming missiles: Ship radar systems will pick up a sea-skimming missile as it comes over the horizon, which at low level may be less than 10 kilometres away, and in rough weather might be less than five.

8 Quick reactions: A missile like Exocet will cover five kilometres in under 20 seconds, so close-in defensive systems must react very fast to threats.

4 Target kill

Phalanx works on a closed-loop system. With conventional anti-aircraft weapons, no correction of aim is assessed until the round or burst of rounds fired has hit or missed the target. Phalanx's radar tracks both the incoming targets and the outgoing bursts of fire. The angular error between the target and the outgoing rounds can be measured, and corrections to the aim can be applied before the burst has reached the target. It requires a very fast reacting fire control system, a rapidly moving gun mount and a very high rate of fire to be effective, but Phalanx has all three. The system uses the 20-mm Vulcan cannon developed for the Century series of supersonic fighters, and fires DU rounds at very high speed. The resulting flat trajectory makes the fire control solution much simpler to calculate.

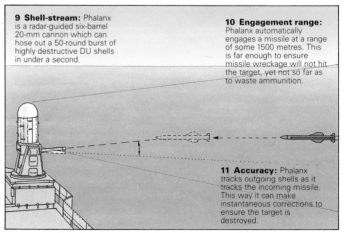

9 Shell-stream: Phalanx is a radar-guided six-barrel 20-mm cannon which can hose out a 50-round burst of highly destructive DU shells in under a second.

10 Engagement range: Phalanx automatically engages a missile at a range of some 1500 metres. This is far enough to ensure missile wreckage will not hit the target, yet not so far as to waste ammunition.

11 Accuracy: Phalanx tracks outgoing shells as it tracks the incoming missile. This way it can make instantaneous corrections to ensure the target is destroyed.

POLARIS
BRITAIN'S LAST RESORT

Specification
Polaris A3TK

Type: submarine-launched ballistic missile
Dimensions: length 9.8 m; diameter 1.4 m
Launch weight: 15875 kg
Propulsion: two-stage solid-propellant rocket
Performance: range 4750 km; CEP 900 m
Guidance: inertial
Warhead: three 60-kiloton-yield Chevaline warheads plus decoys and chaff penetration aids, or three 200-kiloton-yield W-58 warheads

With a genesis reaching back to German plans of World War II, the concept of the missile-armed submarine is hardly new. Both the American and Soviet navies made use of captured German technology in the immediate post-war years, but it was in the 1950s that the concept of the nuclear-powered, nuclear ballistic missile-armed submarine first flowered.

First proposed during the mid-1950s, the Polaris missile was the culmination of one of the most astonishing military, technological and industrial achievements in history. Polaris was a submarine-launched ballistic missile, incorporating a number of brand new concepts never before achieved. These included solid propulsion, lightweight re-entry vehicles, lightweight nuclear and thermonuclear warheads, cold gas-submerged launch, advanced inertial navigation, both for missile and launch platform, and notable quietening of the huge launch submarines.

The programme really got under

A Polaris missile roars out of the ocean during a test launch. It can contain three warheads, each with treble the power of the Hiroshima bomb.

way in 1956, when the US Department of Defense began stating its requirements, and Lockheed began developing a missile to match. The launch of the first Soviet ICBM accelerated developments: the initial (unsuccessful) missile launch took place in 1958, and the first launch of the missile in operational form took place in 1959.

In the meantime, the US Navy was converting five of its nuclear-powered attack submarines on the stocks by the addition of 16 missile tubes, and the first fleet ballistic missile submarine was launched in June 1959. The USS *George Washington* made the first underwater launch of Polaris in July 1960, and in November of that year the submarine and its missiles were declared operational. In the next five years the US Navy was to commission a further 40 boats, establishing a strategic advantage over the Soviets that was to last for two decades.

The initial missile was the UGM-27A Polaris A1. It weighed just over 12 tonnes and had a range of 2250 kilometres, with a single 500-kiloton-yield warhead. By 1959, Lockheed was developing the Polaris A2 (UGM-27B). This had a similar warhead but with range increased to 2750 kilometres. The UGM-27C Polaris A3 missile had its range increased yet again to 4600 kilometres, and in the early 1970s had its single re-entry vehicle replaced by a triple MRV warhead.

UK's Polaris
In 1963, the United Kingdom declared its intention to build four ballistic missile submarines. They were to be armed with the latest Polaris missile, the A3, and were to shoulder the burden of Britain's nuclear deterrent from the Royal Air Force. The 'Resolution' class became

However, as a result of the development of Soviet anti-ballistic missile defences, it was decided in the early 1970s to upgrade the Polaris warheads. The Polaris A3TK Chevaline has its roots in a cancelled US warhead programme called 'Antelope'. Chevaline involved the replacement of the 200-kiloton re-entry vehicles with smaller 60-kiloton devices, each of the re-entry vehicles being hardened against electro-magnetic pulse and hard radiation. The carrier bus was also modified to present defending radar systems with a confusingly large number of credible threats. Although not specified, the modifications are believed to have included the carriage of chaff and decoy penetration aids.

operational in the late 1960s.

The Royal Navy is the last user of Polaris, the US Navy having switched successively to Poseidon C3, Trident D4 and Trident D5 missiles. The British Polaris missiles were given new engines in the 1980s, allowing the force to remain operational until replaced by Trident-equipped boats in the 1990s.

Above: For the crew of a Royal Navy SSBN the objective of every patrol is to remain at sea undetected, ready to fire if so ordered.

British A3 missiles were originally equipped with three British-designed and -built re-entry vehicles, each with a 200-kiloton-yield nuclear device for use against area targets such as cities and oil-fields.

Below: The 'Resolution' class SSBNs have carried Britain's weapon of last resort for over 20 years and will be replaced shortly.

The Royal Navy operates four 'Resolution' class SSBNs, which is the minimum number required to guarantee one on station, ready to fire at all times. Some university 'peace studies' departments claim to understand submarine operations better than the Navy and say that only three Trident-armed boats are required.

SEA CAT

The Sea Cat missile was the first naval SAM designed for close-range air defence. Designed and manufactured by Short Brothers in the late 1950s, it was intended to replace rapid-firing anti-aircraft guns such as the 40-mm Bofors gun. One Sea Cat launcher was supposed to replace a twin 40-mm Bofors mounting.

Developed from the Malkara anti-tank missile, the first Sea Cat systems were command-guided by an operator using a joystick to steer the missile on to the target. Brightly burning flares in the missile's tail help the operator follow the missile's flight. The first trials took place in 1960, and shipboard firings from HMS *Decoy* occurred in 1961. After further shipboard trials the Sea Cat entered service in 1964, known as the GWS (Guided Weapon System) Mk 20 in the Royal Navy.

Sea Cat has a dual-thrust motor, four fixed fins and four hydraulically-driven wings. The first Sea Cats had a warhead containing 17 kilograms of high explosive, whereas Mod 0 Sea Cats have a 14-kilogram continuous rod with a 2.2-kilogram HE charge. Both versions have an impact and infra-red proximity fuse. The missile is controlled by radio command and it can be integrated into almost any sighting and fire-control system. The Royal Navy improved Sea Cat several times before it finally saw action during the Falklands war.

The GWS Mk 21 is controlled by the Type 262 fire-control radar, originally designed for 40-mm anti-aircraft guns. The GWS Mk 22 is also radar-controlled and the visual direction is via closed-circuit television. The latter automatically tracks the missile, which reduces the engagement time since the operator no longer needs to 'gather' the missile in his sights. The addition of radar control enables Sea Cat to engage in darkness or poor visibility.

Export success

Sea Cat proved a popular export, with 17 navies adopting one version or another. A land-based version called Tigercat was developed for export. Argentina bought both types and deployed them during the Falklands war. The most common Sea Cat system in service is a quadruple launcher that is loaded by hand. A lightweight triple launcher is also used.

In 1977 an altimeter was developed that could automatically prevent a Sea Cat missile from straying into the sea. Without the altimeter it was difficult for Sea Cat to engage a target within about 30 metres of the water. Sea-skimming missiles and particularly gutsy pilots attacking at lower alti-

The Short Brothers' Sea Cat surface-to-air missile has been used by the Royal Navy since the 1960s. The first Argentine aircraft attacking San Carlos was brought down by one fired from HMS Plymouth.

Sea Cat is a relatively small missile designed to destroy enemy aircraft at short range. It is relatively slow by modern standards, although it can operate only a few metres above the sea, prevented from hitting the water by its onboard altimeter.

Specification
Sea Cat

Dimensions: length 1.48 m; diameter 190 mm; span 650 mm
Weight: 68 kg
Warhead: 14-kg continuous rod
Performance: speed Mach 0.6 (at end of boost); range 5 km; altitude 6 to 1000 m

Mach 0.6 (approximately 200 m/sec). Since the Argentine pilots excelled at very fast attacks at wave-top height, Sea Cat was simply too slow and too short-ranged to catch them. It certainly spoiled the aim of many attacking aircraft – a warship firing live missiles is a very different proposition to a target on a peacetime training area – but the true aim of a point-defence missile is to shoot down the attacker. A ship attacked by determined pilots could not trust Sea Cat to stop them. The Sea Cat-armed frigate HMS *Plymouth* was attacked on 8 June by five IAI Daggers which managed to hit her with four 1,000-lb bombs despite a succession of missiles (and 20-mm cannon) fired at them.

A faster version of Sea Cat was proposed, but the Royal Navy adopted Seawolf instead. This does offer supersonic performance at sea-skimming altitudes but only at the cost of much greater complexity.

Above: HMS Amazon enters Pearl Harbor during a 1986 exercise. Two missiles are loaded in the Sea Cat system.

Below: HMS Antelope was sunk by Argentine aircraft in the Falklands war. She blew up when the flames reached her Sea Cat magazine.

tudes were impossible to hit. Sea-skimming trials in 1980 proved successful and Royal Navy Sea Cats were modified to attack targets within about 6 metres of the surface.

Nearly 20 years after it had entered service, Sea Cat was finally fired in anger during the Falklands war. Although as many as eight Argentine aircraft were claimed to be destroyed by Sea Cat, only a couple were definite Sea Cat kills. Many of the Argentine aircraft shot down over San Carlos were under fire from Sea Dart, Sea Cat and Seawolf missiles, not to mention anti-aircraft guns and Rapier SAMs from ashore.

Sea Cat showed its age during the Falklands war. Although it could tackle very low-flying aircraft, it is a relatively slow missile travelling at

SEA DART Fleet Defender

Left: A Sea Dart area defence naval SAM blasts clear of the twin-rail launcher on a Royal Navy Type 42 destroyer. Sea Dart was developed as the primary defence for British carrier battle groups that were being planned in the early 1960s.

Above: The Type 42 destroyer was designed in the late 1960s after the new British carrier and its escorts were cancelled. For reasons of economy it was the smallest platform that could take the Sea Dart system. This had the side effect of making it very difficult to upgrade the vessel with new equipment without causing top-heaviness.

The British Aerospace Dynamics Sea Dart missile is now getting on in years, first prototypes having been test-fired in 1965 and entering production in 1968. It has nevertheless been successful in combat in a variety of climates. It coped with both the bitter cold of the South Atlantic winter during the Falklands campaign of 1982 and the warmth of the Persian Gulf during the US-led coalition's 1991 war against Iraq.

Sea Dart is an area defence missile, with a maximum range of more than 40 kilometres. It was designed primarily as an anti-aircraft system, and was used as such in the Falklands, when it displayed considerable versatility in destroying targets ranging from a Lear Jet flying at 50,000 feet to an A-4 Skyhawk skimming the surface of the sea.

Gulf success

Sea Dart has a useful anti-missile performance. This was tested in action during February 1991, when an Iraqi 'Silkworm' anti-ship missile fired at the coalition fleet in the Gulf was destroyed by two Sea Darts launched from the Type 42 destroyer HMS *Gloucester*.

Sea Dart is basically a flying ramjet, with guidance systems, fuses, warheads and fuel tanks all being wrapped around the ramjet's central air duct. It is launched from a twin-rail launcher by means of a solid fuel booster, and is then sustained in flight by the ramjet. Sea Dart's sensitive guidance system, coupled with very fast-acting cruciform control surfaces, mean that the missile is capable of great accuracy.

A greatly improved variant of Sea Dart intended to cope with the latest generation of Soviet missiles was to be built in the 1980s, but fell victim to the short-sighted defence economies typical of Mrs Thatcher's Conservative government.

Specification

GWS30 Sea Dart

Type: medium range area defence naval surface-to-air missile
Dimensions: length 4.36 m; diameter 0.42 m; span 0.91 m
Launch weight: 550 kg
Warhead: HE blast-fragmentation
Performance: maximum speed greater than Mach 3; range 40-60 km; engagement envelope between 30 and 18300 m

Sea Dart is a highly capable area defence weapon that has been combat-tested at low and high levels, against aircraft and missiles.

Engagement sequence

1 Surveillance

As with most air defence systems, the Sea Dart missile is only one part of a complex network of electronics and weaponry. The long-range Type 1022 radar ahead of the foremast scans sea and sky in search of targets. Once a target is detected, the slender Type 992 target indication radar atop the main mast provides more detailed target information to the ship's computerised action information system. Target information and track is now passed to one the two Type 909 radars, their large radomes prominent at the fore and aft ends of the superstructure. These are the systems that actually control the missile engagement. The main surveillance radar has a detection range of over 100 kilometres or more for high-flying targets, and the tracking and fire-control radars are probably effective out to beyond 50 kilometres.

2 Launch

Sea Dart missiles are stored in a fully automatic 22-round magazine below decks, which feeds a twin-rail missile launcher. The ship's ADAWS-4 action information system determines the nature of the threat and selects the most appropriate weapons system to deal with it. The Sea Dart missiles are accelerated by a solid rocket booster motor to the supersonic speed necessary for the main ramjet sustainer motor to work effectively. Travelling at three times the speed of sound, the missile can reach its maximum range and operating height of 65 kilometres and 18000 metres in about a minute. Sea Dart's minimum intercept altitude is quoted as 30 metres, or about 100 ft, but in the Falklands a Sea Dart knocked down an Argentine A-4 Skyhawk skimming less than 20 ft above the ocean.

3 Flight

The computers of the ADAWS-4 action information system take in data from the Type 1022 search radar, the Type 922 tracking radar, and the Type 909 fire control radars. They use the data to compute the position of the target. Missile launchers are automatically pointed along the threat axis, and the Sea Darts will be launched at the optimum moment to effect an intercept as far from the launch ship or the fleet as possible. Sea Dart's ramjet gives it controlled and stable propulsion through its flight, without the continuous acceleration of rocket-powered missiles, making interceptions simpler to calculate. An improvement programme in the late 1980s, while not the radical performance enhancement envisaged earlier in the decade, has nevertheless given Sea Dart improved terminal homing performance, increasing its lethality.

4 Intercept

Sea Dart is a semi-active radar homing (SARH) missile. It has a seeker in the nose that homes in on the energy transmitted from the ship's radar systems and reflected from the target. Once within range, the proximity fuse detects the incoming missile and detonates the warhead. Early models of the Sea Dart had a high-explosive continuous-rod warhead. This flings a long chain of metal, rather like a huge lassoo, into the path of the target. Current Sea Darts have high explosive blast-fragmentation warheads, which send clouds of razor-sharp metal fragments ripping through the target missile. Sea Dart is a highly effective missile, but its magazines and handling gear mean that it can only be launched from ships of 4,000 tons and above. However, British Aerospace have developed a lightweight system for vessels displacing as little as 300 tons.

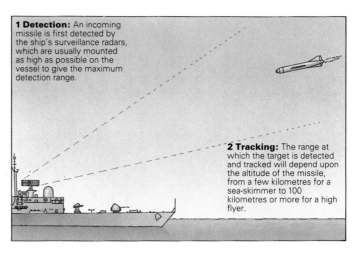

1 Detection: An incoming missile is first detected by the ship's surveillance radars, which are usually mounted as high as possible on the vessel to give the maximum detection range.

2 Tracking: The range at which the target is detected and tracked will depend upon the altitude of the missile, from a few kilometres for a sea-skimmer to 100 kilometres or more for a high flyer.

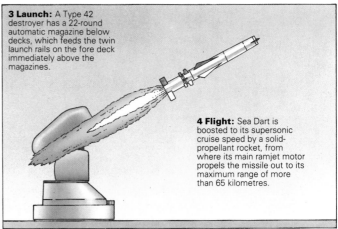

3 Launch: A Type 42 destroyer has a 22-round automatic magazine below decks, which feeds the twin launch rails on the fore deck immediately above the magazines.

4 Flight: Sea Dart is boosted to its supersonic cruise speed by a solid-propellant rocket, from where its main ramjet motor propels the missile out to its maximum range of more than 65 kilometres.

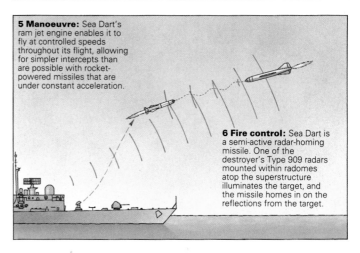

5 Manoeuvre: Sea Dart's ram jet engine enables it to fly at controlled speeds throughout its flight, allowing for simpler intercepts than are possible with rocket-powered missiles that are under constant acceleration.

6 Fire control: Sea Dart is a semi-active radar-homing missile. One of the destroyer's Type 909 radars mounted within radomes atop the superstructure illuminates the target, and the missile homes in on the reflections from the target.

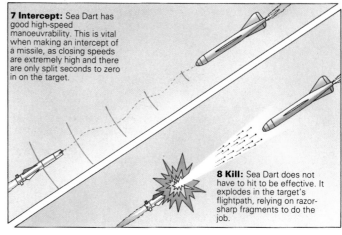

7 Intercept: Sea Dart has good high-speed manoeuvrability. This is vital when making an intercept of a missile, as closing speeds are extremely high and there are only split seconds to zero in on the target.

8 Kill: Sea Dart does not have to hit to be effective. It explodes in the target's flightpath, relying on razor-sharp fragments to do the job.

SEAWOLF

Specification

GWS 25 Seawolf

Dimensions: length 2 m; diameter 18 cm; span 40 cm
Weight: 81 kg
Maximum speed: Mach 2+
Range: 7 km
Engagement envelope: 5-3000 m
Guidance: radio CLOS (Command-to-Line-Of-Sight)
Warhead: 14-kg high-explosive fragmentation
Fuse: proximity and contact

Seawolf is treated like a round of ammunition, needing little preparation before being loaded manually into its six-round launcher.

Nowhere has the post-war technological revolution affected warfare more than at sea. Today's warships fight with radar, computers and missiles, to the extent that where fighting ships formerly bristled with light and heavy anti-aircraft guns, they now are defended in the main by one or two missile launch systems.

British warships are equipped with one of the most effective of all short-range naval air defences, in the shape of the Seawolf system. Surprisingly, this most effective of weapons can trace its origins back three decades to a specification issued in 1962, with development starting in the late 1960s and service use starting 10 years later.

Sea Cat replacement

Seawolf was originally developed as a replacement for the widely fitted Sea Cat SAM system. However, the original GWS 25 Seawolf system was quite heavy, and was not found to be practicable for vessels of less than 2,500 tonnes. In fact, it was the space required to fit two manually-reloaded six-round launchers with their attendant 30-round magazines and radar, launch and command systems which dictated that the 'Type 22' frigate should actually displace more than the 'Type 42' destroyer.

Seawolf is a fully automatic point-defence system, with radio command-to-line-of-sight guidance coupled with radar or low-light TV tracking. Highly manoeuvrable and deadly accurate, Seawolf is capable of intercepting other missiles, and in tests has even scored a direct hit on a 4.5-inch shell in flight.

Seawolf saw its combat debut in the Falklands, and in its first action missiles from HMS *Brilliant* shot down three out of four attacking Argentine A-4 Skyhawks. Some problems arose, however, as the system had difficulty in tracking targets flying close together. These were solved by changing the system's computer software.

Vertical launch variants

Vertical launch variants of the Seawolf have been under development since the late 1960s. Vertically-launched missiles have a number of advantages over those fired from more conventional launchers: they eliminate blind arcs, have shorter reaction times, and the additional thrust vectoring booster motor which gets the missile into the air also gives it increased range. Vertical launch Seawolf has an effective range of about 10 kilometres compared with the seven kilometres' range of earlier variants. The GWS 26 Mod 1 Seawolf system comprises modules of eight canisters, controlled by a missile fire unit. The standard layout is a four-module silo containing 32 missiles. Seawolf VL is shipped aboard the Royal Navy's Type 23 'Duke' class frigates, and is also fitted to the new 'Fort Victoria' class of fleet replenishment ships.

A lightweight Seawolf variant has been developed, using a modified version of the old four-missile Sea Cat launcher. Originally a private venture to make the system more suitable for use aboard smaller ships, the GWS 26 Mod 2 Seawolf has been ordered by the Royal Navy to add to the close-range air defences of 'Batch 3 Type 42' destroyers and the three 'Invincible' class light carriers.

Engagement sequence

1 Incoming

Seawolf is designed to engage fast, low-flying targets. Target detection is performed by the launch platform's surveillance radar. In the original Seawolf system, the surveillance sensor was to be the Marconi Type 965 long-range radar, as fitted to 'Type 42' destroyers. However, the later Type 967/968 air- and surface-search system is more appropriate, and was fitted to the Seawolf-armed 'Type 22' frigates. This is a short-/medium-range system with back-to-back antennas in a single mount rotating once every two seconds. It operates in the D-band of the radar spectrum, and with sophisticated data-handling can calculate target speed and range. The system incorporates both IFF and ECCM to ensure that friendly aircraft are not attacked and that enemy countermeasures can be negated.

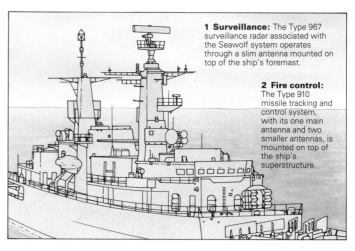

1 Surveillance: The Type 967 surveillance radar associated with the Seawolf system operates through a slim antenna mounted on top of the ship's foremast.

2 Fire control: The Type 910 missile tracking and control system, with its one main antenna and two smaller antennas, is mounted on top of the ship's superstructure.

2 Tracking

Range, bearing and target velocity are passed automatically to a Ferranti fire-control computer. The system will determine within five seconds if the incoming track is hostile. If so, an appropriate launcher with its associated tracker is aligned on the target. The Type 910 tracking radar scans up and down in the designated direction, usually acquiring the target in a fraction of a second. It can lock onto a 30-centimetre diameter target at ranges of 10 kilometres or more. Low-flying targets can sometimes be lost in clutter from the sea, and in that case tracking is automatically switched to a manually-operated TV tracker. Missiles are automatically launched once the target interception point comes within missile range, and a single tracker can simultaneously control three missiles as long as they are aimed at the same target.

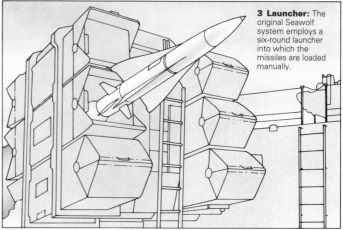

3 Launcher: The original Seawolf system employs a six-round launcher into which the missiles are loaded manually.

3 Command guidance

The Type 910 radar and the succeeding lightweight Type 911 radar possess a wide angle radar beam as well as a narrow target beam and a microwave radio command-guidance beam. Once missiles are launched, they are automatically gathered into the wide angle beam and command-guided onto the line-of-sight. The antenna uses electronic angle tracking to measure small deviations in the missile's track without the need for moving the whole antenna. In TV-guided mode, the system switches to semi-automatic command-to-line-of-sight, in which the target is tracked manually. The operator holds a set of crosshairs on his monitor screen onto the target and the fire-control system then gathers and steers the missile automatically.

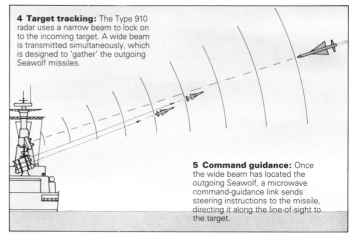

4 Target tracking: The Type 910 radar uses a narrow beam to lock on to the incoming target. A wide beam is transmitted simultaneously, which is designed to 'gather' the outgoing Seawolf missiles.

5 Command guidance: Once the wide beam has located the outgoing Seawolf, a microwave command-guidance link sends steering instructions to the missile, directing it along the line-of-sight to the target.

4 Vertical launch

Seawolf VL is launched from a 32-round silo. Surveillance is provided by the improved Type 996 radar, details of which are still classified but which has much better performance than the original Type 967/968, particularly in difficult radar environments. The tracker system is the Type 911, which is considerably lighter than the original Type 910. It uses advanced signal processing to provide maximum performance in severely cluttered conditions, such as against very low-flying sea-skimming missiles or in shallow land-locked waters. The millimetric guidance radar is adapted from the land-based Rapier Blindfire system and can track outgoing Seawolf missiles even when they are very close to the sea.

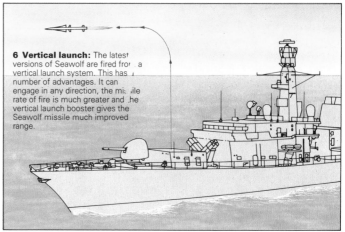

6 Vertical launch: The latest versions of Seawolf are fired from a vertical launch system. This has a number of advantages. It can engage in any direction, the missile rate of fire is much greater and the vertical launch booster gives the Seawolf missile much improved range.

SONAR
Supersense

Sonar is the only effective underwater sense, both for submarine crews (above) listening for enemy attackers and for ASW surface ships like the Soviet navy's 'Moskva' class cruisers (left).

Underwater warfare is unlike any other form of conflict. A duel between submarines is a bit like that between two men, blindfold in a darkened room. Both carry loaded revolvers, but the only clues to each other's position come from the sounds the opponent makes: the rustle of shoes in the carpet or of moving clothes, or at close range the sound of breathing.

Sound is the only sense that is of any real use under water. Light, radio and radar penetrate the ocean poorly or not at all. Sound, on the other hand, can travel great distances. The first underwater sensors developed in the early years of this century were simple hydrophones. By using an array of several such listening devices, submarines could be detected at a distance of thousands of yards, and even general directions might be determined.

During World War II, active sound systems like the British ASDIC and the American Sonar were developed. They used bursts of sound causing echoes to detect targets, in much the same way that radar uses radio waves to detect aircraft. Sonar, which was originally an acronym for SOund Navigation And Ranging, has become the generic term for any underwater

use of echo-location, and has even been applied to the similar techniques used by whales and dolphins.

Hidden subs

Technology has made sonar effective beyond the wildest dreams of early operators, but the sea remains a confusing place. In spite of all the highly-advanced sensor gear and computerised signal-processing equipment that make up a modern sonar system, the performance of that system will vary greatly depending on the conditions. For one thing, sound does not travel in straight lines under water. So much depends upon transient ocean conditions, such as water temperature and salinity, and confusing echoes can be generated by the sea bed, storm waves on the surface, schools of fish or pods of whales.

A basic fact of sonar operations is that the lower the frequency of the sound, the further it will travel. However, there is a trade-off between range and discrimination. Small, high-frequency sonars have limited range but are useful for providing detailed in-

formation in specialist situations, such as mine-hunting or when operating under ice. Lower frequency sonars have longer range but provide less detailed information about targets. In general, the larger the sound generator, the lower the frequency it can generate. This may go some way to explaining why modern submarines and ASW vessels are so much larger than earlier vessels. Anti-submarine destroyers and frigates usually have a large bow sonar – often indicated by a sharply-raked stem, designed to keep the anchors clear of the sonar dome – that is used as an early-warning system, with helicopters being used to close in precisely on the contact using their own higher frequency dipping sonars.

In spite of numerous efforts to develop new sensor systems to detect submarines under water, sound remains the only practical method, and will remain so for the foreseeable future.

Helicopters that are fitted with dipping sonars at the end of long cables can become potent ASW weapons, either independently or as an extension of an ASW vessel's sensor and weapon fit.

Engagement sequence

1 Convergence zones

Sound rarely travels in a straight line in water. It is affected by pressure, temperature, salinity and other factors. A sound beam will tend to bend with depth as its speed is slowed, until it reaches really deep water when the process is reversed and it begins to bend upwards again. These sound waves will concentrate in a 'convergence zone', or CZ, which is a ring between two and 10 kilometres wide. In the North Atlantic this CZ will form about 60 kilometres out from the listening vessel. If the sonar beam is powerful enough, it will go through the process a second or even a third time, forming CZs at 130 and 195 kilometres. Targets can be detected passing through the zones, although such long-range sonar can only indicate their presence, not their positions.

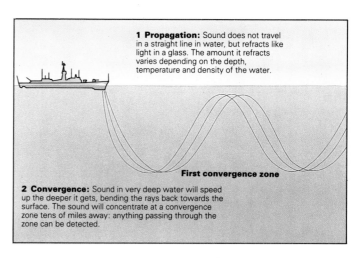

1 Propagation: Sound does not travel in a straight line in water, but refracts like light in a glass. The amount it refracts varies depending on the depth, temperature and density of the water.

2 Convergence: Sound in very deep water will speed up the deeper it gets, bending the rays back towards the surface. The sound will concentrate at a convergence zone tens of miles away: anything passing through the zone can be detected.

First convergence zone

2 Surface layer

The speed of sound in water is determined by a number of factors, but temperature is one of the most important. Since water cools as it gets deeper, the speed of sound varies with depth. However, the surface layer of the sea is generally kept at nearly uniform temperature by the action of wind and waves. A constant temperature means constant sound velocity, and sound waves will travel in reasonably straight lines. In such cases, sonar works pretty much like radar, although maximum range in ideal conditions is about 20000 metres, and is often much less. The surface layer can vary considerably in thickness: in the cool, rough waters of the North Atlantic it can stretch as far down as 600 ft, while in the Mediterranean in summer it might be less than 50 ft thick.

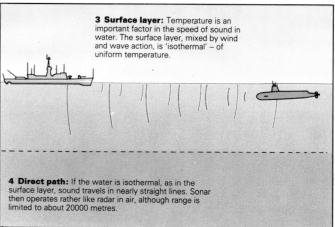

3 Surface layer: Temperature is an important factor in the speed of sound in water. The surface layer, mixed by wind and wave action, is 'isothermal' – of uniform temperature.

4 Direct path: If the water is isothermal, as in the surface layer, sound travels in nearly straight lines. Sonar then operates rather like radar in air, although range is limited to about 20000 metres.

3 Bounce modes

There is a discontinuity between the isothermal surface layer and the normal waters of the ocean beneath. Under certain circumstances and at certain angles of transmission, a sonar beam can be bounced off the layer, and if the sea is not too rough, it can also bounce off the surface. In this way, the surface layer acts as a wave guide. Trapped in the layer, a sound beam can travel longer distances than would otherwise be possible. Sound can also be bounced off the sea bottom. With an estimated maximum range under ideal conditions of some 40000 metres, bottom bounce was thought, in the 1950s, to promise to fill the gap between direct path sonar and the convergence zones. Unfortunately, hydrographic surveys have revealed that much of the world's sea bed is more absorbent than expected, and ideal conditions rarely occur.

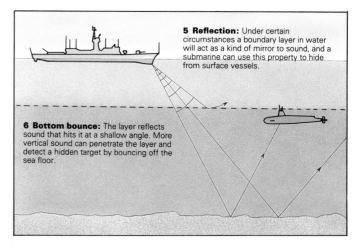

5 Reflection: Under certain circumstances a boundary layer in water will act as a kind of mirror to sound, and a submarine can use this property to hide from surface vessels.

6 Bottom bounce: The layer reflects sound that hits it at a shallow angle. More vertical sound can penetrate the layer and detect a hidden target by bouncing off the sea floor.

4 Towed and variable depth sonars

The most effective sonar systems currently in service are towed arrays, which are pulled along behind a vessel at the end of a long cable. These have a number of advantages. Placing the sonar some distance from the ship or the submarine cuts out the problems of self-generated noise. By using a number of transponders in a long linear array, the system can operate like one very large system, effective in listening for very low frequencies at maximum range. By varying the depth of the sonar, lowering it beneath the layer, a surface ship can also hunt for submarines hiding behind the reflective barrier, or listen for long-range CZ sounds. Helicopters can be equipped with dipping sonars, which while less powerful than ship-mounted sets, can also probe beneath the surface layer.

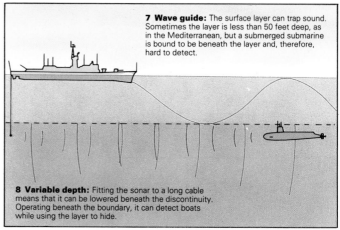

7 Wave guide: The surface layer can trap sound. Sometimes the layer is less than 50 feet deep, as in the Mediterranean, but a submerged submarine is bound to be beneath the layer and, therefore, hard to detect.

8 Variable depth: Fitting the sonar to a long cable means that it can be lowered beneath the discontinuity. Operating beneath the boundary, it can detect boats while using the layer to hide.

The launching bay underneath an S-3 Viking contains a full range of sonobuoys to be dropped around suspected submarine contacts.

SONOBUOYS

In many ways, the maritime patrol aircraft is the perfect weapon for anti-submarine warfare. Faster than surface ships, it can cover vast areas of ocean in a relatively short time. As long as it can find the enemy, it can attack a submerged boat with little fear of counterattack. Unfortunately, finding the submarine is the problem.

Anti-submarine warfare is a difficult business. In every other military field you can rely on visual, radar or infra-red techniques to detect and home in on targets, but under the sea, except for some special cases, you have to rely on sound. And sound can behave in very strange ways. That is not the real problem, however. It is impossible for an aircraft flying at high speed through the air to detect sound through the barrier of the ocean's surface. It cannot dip a hydrophone into the water like a helicopter, or a surface ship, or a submarine. Without some kind of sound-sensing device, it has to rely on MAD – magnetic anomaly detection – gear, or 'sniffers', or radar.

MAD systems detect the magnetic signature of a submerged submarine. They are very accurate, but short-ranged, and are not particularly effective against deeply submerged submarines. Sniffers detect the diesel exhaust from a snorkelling submarine. Nuclear submarines produce no exhaust gases, so they don't need snorkels. Neither do the latest conventionally-powered submarines, equipped as they are with closed-cycle engines. Radar can be used to detect a submarine's periscope. But if a submarine commander elects to operate passively, using his sonar alone, then there is no periscope breaking the surface for an ASW aircraft's radar to detect.

Expendable systems

The solution to the problem is the sonobuoy. Sonobuoys are expendable acoustic systems, ejected from the aircraft as it flies over the surface of the ocean. They are generally of two parts: a sensor section, which can be pre-set to descend to a specific operating depth, and a radio transponder, which remains afloat on the surface.

The sensor can be active or passive, the passive unit listening for the sound of enemy submarines, while the active version generates its own sound waves, which hopefully will bounce off the target and back to the sonobuoy. Data is transmitted along a wire connecting the submerged portion of the equipment to the floating component, which transmits its information back to the launching aircraft.

An ASW aircraft will normally lay a pattern of six to 10 buoys, laying out a sound net in the hope of catching an enemy submarine. Careful analysis of the signals from several sonobuoys should give an indication of the submarine's position, course and speed. The earliest equipment was relatively short-ranged, and had no direction-finding ability, but constant development has seen vast improvements.

The introduction into service of LOFAR (low frequency) technology has given small sonar sets great range. Directional LOFAR buoys, known as DIFAR, can, as their name suggests, pick out the direction of a target. Two DIFAR buoys are enough to get a fix on the target by cross bearings.

Active buoys are used to 'ping' the target. However, once they are used, a hostile submarine can hear the sound transmissions and will know that there is a hunter aircraft about. It is also heavy on battery power, and an active buoy has only a few hours of life at the most.

Most active buoys are command activated by the ASW aircraft. Until that time, they operate passively. They will only come into play in the final stage of an attack, in order to fix a target submarine's position exactly. The data is fed into the aircraft's fire-control system, which will then compute the exact moment and position to drop homing torpedoes or, more rarely, depth charges.

A US Navy S-3 Viking flies over a surfaced US submarine. The Soviets were so concerned about sonobuoy-dropping aircraft that they experimented with SAMs fired from submarine conning towers.

Engagement sequence

1 Ejection

An anti-submarine warfare aircraft needs to be able to carry and eject large numbers of standard-sized sonobuoys. Typically, a long-range patrol plane such as the US Navy's P-3 Orion has 48 pre-loaded rearward-slanting launchers under the cabin, each of which contains a NATO standard 'A' sized buoy, 124 mm in diameter. Thirty-six more buoys are stored in racks in the aircraft's cabin, which can be loaded in flight into three more launch tubes which are accessible to the crew. Other aircraft, such as the RAF's Nimrod MR.Mk 2s, carry all of their buoys internally, launching via pressurised tubes or rotary launchers.

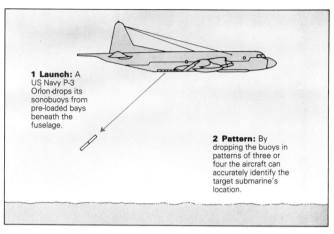

1 Launch: A US Navy P-3 Orion drops its sonobuoys from pre-loaded bays beneath the fuselage.

2 Pattern: By dropping the buoys in patterns of three or four the aircraft can accurately identify the target submarine's location.

2 Passive buoys

The US Navy's AN/SSQ-53B is its primary sonobuoy sensor. It can be dropped by aircraft travelling at up to 325 knots and from as high as 40,000 ft. It is slowed by parachute, which is jettisoned as it reaches the surface. Sea water activates the battery system, which separates the sonar sensor from the communication buoy. The SSQ-53B can be programmed to sink to 30, 120 or 300 metres, enabling it to penetrate the surface layer of the water whatever the weather and sea conditions. SSQ-53 is a DIFAR – direction finding and recording – sensor, listening for low-frequency sounds and indicating their direction.

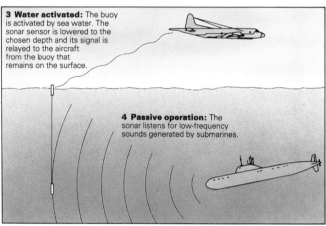

3 Water activated: The buoy is activated by sea water. The sonar sensor is lowered to the chosen depth and its signal is relayed to the aircraft from the buoy that remains on the surface.

4 Passive operation: The sonar listens for low-frequency sounds generated by submarines.

3 Active buoys

The SSQ-963 Command Active Multi-Beam sonobuoy is a typical advanced active system currently in service with the Royal Air Force. It combines the capacity of a direction-finding passive sonobuoy, able to hear snorkelling diesel submarines at ranges of more than 50 kilometres, with a transmitting acoustic projector. Coded radio commands from the launch aircraft are necessary before the buoy can go active. It is also more flexible than earlier systems, which had hydrophones that were pre-set to descend to a specified depth. The SSQ-963's depth is adjustable by radio command from the launch aircraft.

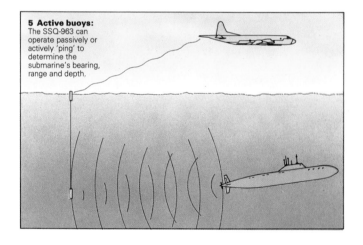

5 Active buoys: The SSQ-963 can operate passively or actively 'ping' to determine the submarine's bearing, range and depth.

4 Measurement and communication

Sonobuoys are not the only air-launched sound equipment. Aircraft are at a disadvantage when they do not know the water conditions below the surface. Bathymetric buoys are like sonobuoys, except that they sink slowly, taking information on water temperature, pressure and density, as well as sampling ambient noise levels. Communication buoys are the other main type of equipment. They float on the surface, receiving radio signals from the launch aircraft and retransmitting them as sound signals in order to contact a submerged friendly submarine.

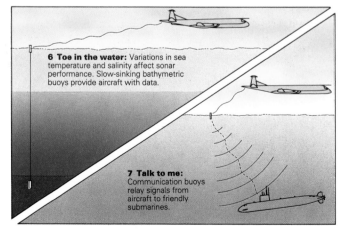

6 Toe in the water: Variations in sea temperature and salinity affect sonar performance. Slow-sinking bathymetric buoys provide aircraft with data.

7 Talk to me: Communication buoys relay signals from aircraft to friendly submarines.

SOVIET ATTACK SUBMARINES

*The **USSR** built 10 4,000-ton 'Charlie I' class nuclear submarines between 1967 and 1972. They have eight SS-N-7 subsonic surface-to-surface missiles with a range of about 60 kilometres and a 200-kiloton warhead.*

Nuclear-powered attack submarines (SSNs) are among the most powerful warships in the modern world. Fast, heavily-armed and difficult to detect, only five navies have the money and expertise to operate them. In 1991, France and China had four operational SSNs each; the Royal Navy had 14; the USSR 60; and the USA 85. This enormous force is a direct product of the Cold War. From the late 1950s the Soviet navy expanded its submarine fleet at a breakneck pace until it posed a serious menace to NATO's vital Atlantic lifeline. If it had ever come to war between the Warsaw Pact and NATO, European fortunes would have depended on seaborne reinforcements of men and material from the USA.

Soviet SSNs are referred to in NATO by letters from the NATO phonetic alphabet, whereas in the Soviet navy they have a number preceded by the letter K. Some also have names: according to *Soviet News* one 'Victor' class submarine apparently rejoiced in the name '60 years of the Young Communist League'! The first Soviet SSNs are described in the West as the 'November' class and 14 were built from 1958 to 1963. Capable of 30 knots submerged and able to dive to nearly 500 metres, they were a dramatic improvement over World War II-era diesel-electric boats. Armed with 533-mm torpedoes, some with 15-kiloton nuclear warheads, they were intended to attack NATO shipping in the Atlantic.

The 'Novembers' were in service for the entire history of the Soviet SSN fleet. One sank in an accident off Britain in 1970, two were disposed of during the 1980s and the 11 survivors were scrapped in 1990-91. They earned a sinister reputation in their 30 years' service. Dubbed 'widow-makers' in the Soviet navy, their reactor shielding was notoriously unreliable and special hospital wards were created to care for sailors suffering from radiation sickness.

The 'Novembers' were followed by the 'Echo' class cruise missile submarines (SSGNs) that carried six SS-N-3 'Shaddock' missiles. With a range of 460 kilometres and carrying either a ton of high explosive or a 350-kiloton nuclear warhead, the SS-N-3 was apparently intended to destroy the US Navy aircraft carriers, as they represented the greatest threat to Soviet submarine operations. However, the SS-N-3 had severe drawbacks. It was not accurate enough to attack a ship 460 kilometres away, and mid-course guidance was required either from the launching submarine or from a third

Soviet nuclear attack submarines have a poor safety record. In 1970 this 'November' class boat sank in the Atlantic, south west of the UK, after a fire on board.

party in contact with the target. The submarine had to surface to launch the missiles, which took 20-30 minutes to fire, and it could not submerge until the missiles received mid-course guidance. It was consequently very vulnerable.

The development of the SS-N-7 anti-ship missile that could be fired from underwater gave Soviet submariners a much more realistic chance of taking on a NATO surface task force. From 1967 to 1980 the Soviet navy built some 16 'Charlie' class SSGNs fitted with the SS-N-7 and later the SS-N-9. Using active radar homing and with ranges of 35 and 60 kilometres, respectively, these offered the prospect of deadly 'pop-up' attacks on NATO shipping.

Fleet expands

Meanwhile the SSN fleet continued to expand with the 'Victor' class, which began construction in 1965. An improved version appeared in 1972, the 'Victor II', and with the development of the 'Victor III' in 1978, the Soviets narrowed the yawning technological gap between their submarines and those of the US and Royal Navies. Each 'Victor' sub-class introduced better sonar and fire-control systems. The 'Victor IIIs' were much quieter than previous Soviet boats, thanks to anechoic tiling that absorbs sonar emissions and improved propellers built with ultra-modern machine tooling bought from Japan. Armed with SS-N-15 missiles that can deliver a homing torpedo up to 37 kilometres away, the 'Victors' were also a serious threat to NATO submarines. In 1991 there were 25 'Victor IIIs' in service and 22 'Victor Is' and 'Victor IIs'.

From the late 1970s the Soviets launched a series of innovative SSNs. The small 'Alfa' class were the world's fastest, deepest-diving SSNs and caused a sensation in the West when their capabilities became

known. However, the first 'Alfa' commissioned in 1970 was scrapped after only four years. Six more were built from 1979-83 and one of those was scrapped in 1989. Their titanium hulls make them harder to detect and allow them to operate at great depths, but it is a material that is notoriously difficult to weld. They are powered by reactors cooled by liquid metal – a system rejected by the US Navy as too complex and unreliable. Although they are so fast that they can outrun NATO torpedoes, the 'Alfas' are noisy boats and reportedly difficult to maintain.

Fire on board!

A one-off SSN design, designated the 'Mike' class, was identified in the mid-1980s, but it sank off Norway after a fire in 1989. A development of this type, designated the 'Sierra' class, has since entered service and three or

four boats were operational by 1991. Another SSN class, the 'Akula', was also beginning series production in the 1980s, with at least seven boats completed and six more under construction. The 'Akulas' are quiet and capable SSNs probably intended to succeed the 'Victors'.

The Soviet nuclear attack submarine fleet improved enormously during the 1980s, and NATO warships found Soviet submarines harder to detect. The Soviets' own sonar systems improved, and they presented a genuine menace to NATO's Atlantic supply lines. However, while their advanced weapon systems and technical performance tended to attract great attention in the West, the true state of the fleet has only emerged with the downfall of the communist regime.

The Soviet submarine fleet had a terrible safety record; at least three

Above: Soviet officers observe from the conning tower of a 'Victor II' class attack submarine. All seven 'Victor IIs' were based with the Northern Fleet.

Below: The 14,000-ton 'Oscar' class SSGNs carry 24 'Shipwreck' radar-homing anti-ship missiles with a range of 450 kilometres.

Soviet attack submarines 1958-92

The Soviet navy expanded at breakneck speed during the 1950s and 1960s, with a very ambitious programme of submarine construction. It was recognised that if the USSR went to war with the NATO powers, control of the Atlantic Ocean would be critical. The huge Soviet submarine fleet was built to cut off Europe from reinforcements and supplies from America.

'November' class
Displacement: 4,200 tons
Number built: 14
Main armament: guided 533-mm torpedoes with conventional or nuclear warheads

The 'November' class had numerous problems with their reactors and many crew suffered radiation injuries. Some were reportedly kept in service when unsafe at the orders of communist party officials.

'Echo II' class
Displacement: 5,000 tons
Number built: 29
Main armament: eight SS-N-3 anti-ship cruise missiles

The 'Echo II' class nuclear submarines had to surface to fire their missiles. These inaccurate SS-N-3s relied on nuclear warheads to make up for their lack of accuracy.

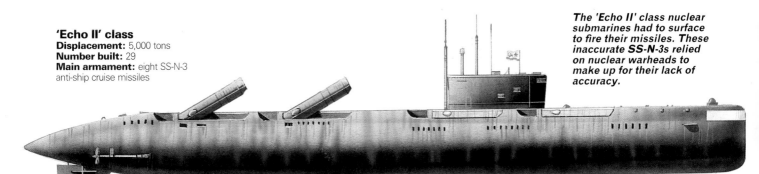

'Alfa' class
Displacement: 2,800 tons
Number built: 7
Main armament: SS-N-15 anti-submarine nuclear missiles, plus guided torpedoes

The innovative 'Alfa' class are still the world's deepest-diving warships and are so fast they can outrun torpedoes. However, they are difficult to maintain and may be withdrawn from service with the new Russian navy soon.

'Victor III' class
Displacement: 5,000 tons
Number built: 26
Main armament: SS-N-21 cruise missiles, plus anti-submarine nuclear missiles and guided torpedoes

The 'Victor III' class SSNs represented a major advance over earlier Soviet nuclear submarines. Harder to detect and fitted with more accurate weapons, they narrowed the technical gap between the Soviet fleet and NATO navies.

Soviet SSNs sank after catching fire. Reports in the Soviet press have highlighted the lack of firefighting equipment and the low level of training throughout the navy. Behind the facade of superpower strength, the Soviet SSN fleet has always been handicapped by a number of problems widespread in the old USSR. The quality of work in the navy yards was generally poor, with several nuclear submarines not accepted for service because the workmanship was so bad. The supply of spare parts was erratic and their quality was often inferior. Technical support at many naval bases was also limited, and the tendency of the Soviet electrical grid to have power failures and voltage surges played havoc with the advanced equipment aboard the submarines. So many naval nuclear reactors failed that at least a dozen have been reportedly dumped in the sea at the edge of the arctic ice cap off Novoya Zemlya.

The US destroyer USS Peterson passes a Soviet 'Moma' class survey ship and a surfaced 'Victor III'. Note the sailors on the submarine's hull.

more nuclear submarines, but the shortages of spare parts and erratic deliveries of fuel began to seriously curtail operations. The policy of keeping submarines in service even when old and unsafe contributed to the high accident rate. The following year the communist coup against President Gorbachev failed and the USSR began to rapidly disintegrate. The nuclear submarine fleet has been inherited by Russia, but whether it has the resources to maintain the fleet remains to be seen.

Right: By 1991 the Soviet fleet had three 'Sierra' class submarines in service. Like the 'Alfas' these are deep-diving and very fast, using a liquid metal-cooled reactor. This is the first 'Sierra' to enter service, seen in 1984.

Below: One of the five remaining 'Alfas' underway on the surface. With their speed and great diving depth they were intended to hunt NATO submarines.

Crew quality

Bad seamanship also contributed to Soviet naval disasters. It was not unusual for a third of the conscripts sent on board a submarine to be unable to speak Russian. The policy of keeping every ship ready for the first day of the war did not help either. Soviet submarines, like their surface fleet, spent much of their time at anchor so they could be guaranteed operational when World War III began. This placed less wear on the warships but did not produce skilled submarine crews.

In 1990, the USSR launched six

SOVIET BALLISTIC MISSILE SUBMARINES

Tom Clancy's novel The Hunt for Red October *introduced the Soviet navy's most powerful submarine to the world. Until the late 1980s the terrible danger posed by these monstrous submarines was known only to naval personnel and defence specialists.* The 'Typhoon' class ballistic missile submarines displace 21,500 tons – more than the Royal Navy's 'Invincible' class aircraft carriers! They are the largest submarines ever constructed, and a single one of them could lay waste to an entire country. Each one carries 20 SS-N-20 nuclear missiles, which have up to nine 100-kiloton warheads. By comparison, the atomic bomb that destroyed Hiroshima in 1945 was rated at 20 kilotons.

The SS-N-20 is credited with a range of 8300 kilometres, enabling Soviet submarines to lurk in their home waters on the edge of the Arctic icepack and still be able to hit American cities.

By 1992 the fleet had commissioned six 'Typhoon' class submarines, which serve alongside over 40 'Delta' class ballistic missile submarines (SSBNs) equipped with equally long-

A US Defense Department illustration shows how the Soviet navy planned to fight and win a sustained nuclear war. Here a 'Typhoon' class SSBN hoists in a second load of ballistic missiles.

ranged weapons. Difficult to locate, powerful enough to wipe out city after city, the Soviet missile fleet was built to fight and win a nuclear war with the Western powers. If the idea of 'winning' such a conflict seems bizarre, the Soviet navy took it very seriously. By the early 1980s the Northern Fleet was practising resupplying SSBNs that had fired off all their missiles. They would rendezvous with special tenders that could reload their tubes in about 24 hours. Meanwhile, Soviet anti-submarine forces ringed the area to keep back NATO attack sub-

The Soviet navy was operating 14 'Delta III' class SSBNs (nuclear-powered ballistic missile submarines) when the USSR collapsed in 1991. Now inherited by the Russian Republic, these enormously powerful warships continue to cruise beneath the Polar ice cap, their missiles still targetted on Europe and the USA.

marines. Until the collapse of the Communist coup d'état in 1991 the Soviet SSBNs were training to fight a sustained nuclear war.

The Soviet missile submarine fleet was very much the creation of

Admiral Sergei Gorshkov. Appointed naval Commander-in-Chief by President Khrushchev in 1956, this dour Cold Warrior served for over 30 years, and by the 1980s he presided over a massive fleet that was steadily narrowing the technological gap between it and the US Navy.

Like the US fleet, the Soviets began their work on missile-firing submarines where the German navy left off in 1944. In 1955 the Soviets successfully fired a naval version of the now infamous 'Scud' missile from a converted 'Zulu' class diesel-electric submarine. Five more modified 'Zulus' then entered service, followed by 23 'Golf' class boats, also conventionally powered and having to surface to launch their three missiles. Accuracy was poor, so large nuclear warheads were fitted in compensation. From 1958 to 1962 the first Soviet nuclear-powered ballistic missile submarines entered service. The 'Hotel' class carried just three missiles, the tubes positioned in an extended fin.

Unfortunately for the Soviet missile fleet, the United States had meanwhile embarked on a larger and much more advanced construction programme. From 1959 to 1964 American shipyards completed 41 nuclear-powered ballistic missile submarines equipped with Polaris SLBMs (submarine-launched ballistic missiles). America's superior industry earned a quantitative and qualitative lead that left the USSR trying to catch up for the next 20 years.

Soviet spying

The USSR tried hard. In the finest Soviet tradition, spies acquired the plans of the US 'Ethan Allen' class missile submarines and key data relating to British long-range sonar. The result was the 'Yankee' class SSBNs, and from 1967 to 1974 the USSR completed 34 of them. They carried 16 SS-N-6 missiles, which could be launched while submerged, and their range of 2400 kilometres allowed boats stationed off the US eastern coast to threaten American cities as far west as the Mississippi. Twelve 'Yankees' remained in service in 1991: 21 of them were disarmed under the terms of the SALT Treaty which limited the numbers of SSBNs and SLBMs to 62 and 950, respectively. In October 1986 the 'Yankee' class submarine K-219 sank some 600 miles north east of Bermuda after an explosion in one of the missile tubes. Four crewmen died.

The 'Yankee' class submarines had to sail long distances across the Atlantic or Pacific to reach their firing positions; en route they were clearly vulnerable to the NATO navies. From 1969 they introduced the SS-N-8,

which had a range of 7800 kilometres, increased to 9100 kilometres in 1977. The 'Yankee' design was enlarged to become the 'Delta' class: giant boats that were the largest submarines in the world in 1972. Eighteen 'Delta I' boats were followed by four 'Delta IIs', which were lengthened to accommodate four more missiles.

From 1976 the Soviets built 14 'Delta IIIs'. These have a distinct 'humped-back' appearance as their hulls have been modified to carry the longer SS-N-18 missiles. The 'Delta IIIs' also represented a substantial advance over earlier vessels. By the

The 'Typhoon' class SSBNs are the world's largest submarines. Two separate 8.5-metre diameter pressure hulls are contained within a free-flooding outer casing, giving a beam of nearly 23 metres. These 21,000-ton boats can punch up through about 3 metres of Arctic ice.

mid-1970s the Soviet spies had succeeded again in delivering vital Western technology. Propulsion systems became quieter and anechoic coatings applied to the hull made these submarines much harder to detect by sonar. With propellers ground on ultra-modern Japanese machine tooling, the Soviet submarines had closed

Soviet ballistic missile submarines

In 1959 the Soviet navy commissioned its first nuclear-powered
ballistic missile submarines (SSBNs). By 1991 it had over 50
SSBNs in service a nuclear strike force capable of devastating
North America and Europe at the touch of a button.

'Typhoon' class
Displacement: 21,000 tons
Number built: 6
Main armament: 20 SS-N-20
missiles (range 8300 km), plus
SS-N-15 and SS-N-16 anti-
submarine missiles and guided
anti-submarine torpedoes

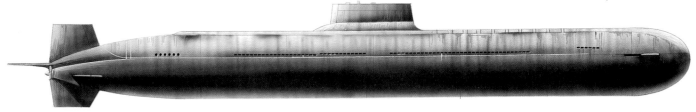

*The first 'Typhoon' class submarine
entered service in 1982 and captured the
imagination of writer Tom Clancy, whose
novel (now a film) revolved around a
'Typhoon' trying to defect to the West. A
single 'Typhoon' carries 180 warheads,
each five times more powerful than the
Hiroshima bomb.*

'Delta III' class
Displacement: 10,000 tons
Number built: 14
Main armament: 16 SS-N-18
missiles (range 8000 km), plus
guided anti-submarine
torpedoes

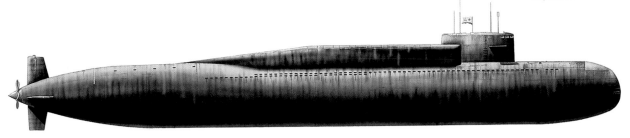

*The 'Delta III' class have the most
extreme 'hump back' profile, as their
missile compartments have been
extended and raised to accommodate 16
SS-N-18 missiles. They were still being
encountered in the North Atlantic in 1991
and are still operating in the Russian
Pacific Fleet.*

'Delta I' class
Displacement: 8,750 tons
Number built: 18
Main armament: 12 SS-N-8
missiles (range 4200 km, later
4900 km), plus guided anti-
submarine torpedoes

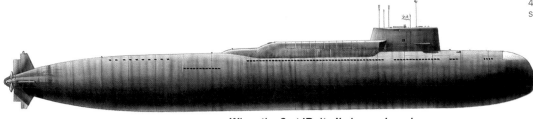

*When the first 'Delta I' class submarine
was completed in 1972 it was the largest
submarine in the world. Its SS-N-8
missiles had over twice the range of
those aboard the earlier 'Yankee' class,
although their sheer size reduced the
number carried to 12. The 'Delta Is'
were built quickly; all 18 were
completed by 1977.*

*The first 'Yankee' class SSBN was built
between 1963 and 1967, and by 1970 the
USSR was building eight of them a year.
To comply with the limitations of the
SALT Treaty some had their missile
tubes removed from 1978 onwards.
Several were converted to fire cruise
missiles, but this programme ground to a
halt in 1991. At least 10 'Yankees' were
still serving as SSBNs at that date.*

'Yankee' class
Displacement: 7,700 tons
Number built: 34
Main armament: 16 SS-N-6
missiles (range 2400 km), plus
guided anti-submarine
torpedoes

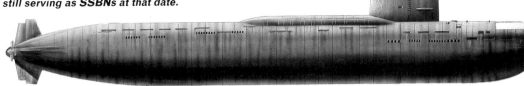

had their share of disasters at sea. One of the 'Golf' class sank in the Pacific and parts were raised by Howard Hughes' salvage vessel *Glomar Explorer* on behalf of the CIA. The 'Hotel' class submarine K-19 suffered a catastrophic coolant leak from its ageing reactor in 1991, killing 10 sailors. Missile tubes have been a particular problem, according to recent reports in the Russian press. Fires and explosions in missile tubes led to at least six major accidents, including the accidental ejection of a live nuclear warhead. The 'Yankee' class boat K-219 suffered a missile tube accident in 1973, 13 years before the second explosion that sank it. As recently as September 1991 a 'Typhoon' class submarine accidentally fired a test missile while in the White Sea.

Frequent collisions

Collisions between the giant Soviet nuclear submarines and NATO boats venturing into their patrol areas have also occurred far more frequently than has been reported. It was reportedly one of the giant 'Typhoon' class that ran into HMS *Splendid*'s towed array when the British submarine was lurking off Murmansk.

The future of the Soviet ballistic missile submarines remains in doubt. Whether the Russian navy which has inherited this massive force has the resources to maintain it, yet alone continue the construction of new vessels, must be open to question.

the technological gap considerably.

These later boats were also much better placed to survive a shooting match with NATO forces. They no longer had to venture far into the Atlantic where NATO's patrol aircraft, underwater surveillance systems and numerous anti-submarine warships were waiting. The submarines of the Northern Fleet could fire from their home ports around Murmansk and still hit the USA. In fact, they trained to operate under the Arctic ice, punching their way through the ice floes before firing. The brute strength of these colossal submarines allows them to batter their way to the surface in a way Western submarines cannot.

Underneath the ice cap

Lurking in their dark and icy lair, the Soviet missile submarines would have been very difficult for NATO to deal with. Sonar conditions under the icepack are dreadful. The moving ice

The 'Golf' class diesel-electric boats had three missiles in an extended fin. A 'Golf II' sank with all hands in the Pacific in 1968. Parts of it were recovered by the CIA.

creates background noise that masks that of loitering submarines. The Soviet navy ringed its so-called 'bastions' with minefields and attack submarines like 'Victor IIIs' that wait in ambush for NATO submarines trying to penetrate the missile submarines' sanctuary. They 'ice-pick' – rising gently against the ice above until its buoyancy is perfectly adjusted and the submarine is wedged into the ice – and they are then impossible to locate with sonar.

Like the nuclear-powered attack submarines, the Soviet SSBNs have

Below and right: On 6 October 1986 the 'Yankee' class SSBN K-219 suffered an explosion in a missile tube while in the Atlantic, 600 miles north east of Bermuda. Four sailors died and the submarine later sank.

SPEARFISH

Designed to meet the Royal Navy's Naval Staff Requirement NSR 7525, the Spearfish is one of the most advanced underwater weapons in the world. Wire-guided, and with dual anti-submarine and anti-ship capability, it is intended to counter the newest generations of fast, deep-diving Soviet submarines.

Spearfish is powered by a liquid-fuelled Sundstrand 21TP01 gas-turbine engine driving a pump jet propulsor. The pump jet was designed incorporating the lessons learned from the highly successful Sting Ray lightweight torpedo programme. It is notably quiet in operation, to help the submarine avoid detection when the torpedo is launched. Selectable high- and low-power outputs enable an attacker to choose maximum speed or maximum range.

Missile control

Spearfish communicates with the submarine via a wire link. This is reeled out of both the weapon and the boat's torpedo tube, which allows high speed launches and enables the submarine to turn away after firing.

The torpedo is autonomous, but it can also be controlled from the submarine. The primary guidance mode is passive, using advanced sensor arrays and transducers to home the torpedo into a noise source. Against very quiet targets, and in the final phases of an attack, Spearfish is equipped with a powerful active sonar. This enables the weapon to home in with deadly precision.

Spearfish can operate in a number of modes such as search, detection and attack. The torpedo's on-board computer controls enable it to classify signals, choose the appropriate attack pattern including re-attacks if needed, and overcome countermeasures and decoys.

Explosive warhead

The warhead is a contact-fused, high-explosive directed-energy or shaped-charge device, intended to penetrate the extremely tough double hulls typical of the largest and fastest Soviet missile and attack boats. Against surface vessels, Spearfish is fused to explode beneath the target ship's keel, thus breaking its back.

The introduction of Spearfish has greatly increased the capabilities of British submarines. Spearfish has roughly twice the speed and range of the preceding Tigerfish design, and its advanced electronics are better.

HMS Conqueror *returns in triumph: but to sink the* Belgrano *she had to use torpedoes designed 50 years ago as her modern ones were too light. Spearfish solves the problem.*

Specification

Spearfish

Type: dual-purpose anti-submarine/anti-ship heavyweight torpedoes
Dimensions: length 6 m; diameter 533 mm
Weight: 1996 kg

Propulsion: HAP-Otto fuel gas-turbine engine driving pump jet
Performance: speed 25 knots (47 km/h) or maximum 70 knots (130 km/h); range 50 km+ (low speed) or 30 km (high speed)
Warhead: 250 kg high-explosive shaped-charge
Fuse: contact or proximity

Below: Taking a swift look at a surface target. Spearfish is capable of destroying enemy surface ships as well as submarines. It can be guided from the launching submarine or be left to home in on the target with its own sonar.

Above: Soviet industry might have trouble building fridges, but it does produce very capable submarines. Spearfish was designed to engage fast, deep-diving targets like this Akula *class attack submarine. Spearfish needs its top speed of over 70 knots if it is to defeat the latest generation of warships.*

Engagement sequence

1 Detection

A submarine's primary sensor is its sonar. Modern submarines have extremely sensitive sound detection devices, which are integrated into some of the most advanced computerised signal-processing suites in service today. These can take a sound signal, filter out the majority of extraneous noise, and identify the source of the sound in a matter of seconds. Sonars aboard a submarine are usually of three forms. The main attack sensors are concentrated in the bow, while a lateral array down the side of the boat acts as a secondary system. The most sensitive sonar, used for surveillance, is the towed array. This streams the listening devices on the end of a long cable behind the boat, taking them clear of all the sounds generated by the normal operations of a submarine, such as engine noises.

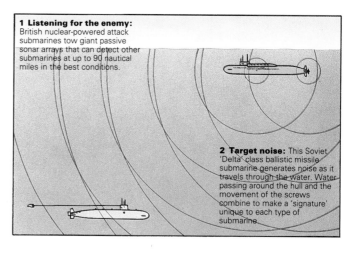

1 Listening for the enemy: British nuclear-powered attack submarines tow giant passive sonar arrays that can detect other submarines at up to 90 nautical miles in the best conditions.

2 Target noise: This Soviet 'Delta' class ballistic missile submarine generates noise as it travels through the water. Water passing around the hull and the movement of the screws combine to make a 'signature' unique to each type of submarine.

2 Passive

Once Spearfish is launched it can accept target data from the mother boat's own on-board sonar, but its main strength comes from the fact that it is capable of autonomous operation. Much of the on-board computer technology was developed for the Marconi Sting Ray lightweight torpedo system, in service with the Royal Navy since the mid-1980s. Advanced signal processors enable the torpedo to pick up all but the very quietest targets, and the on-board computer gives the weapon the ability to make its own tactical decisions during an engagement, varying the direction of the attack and going to an enemy submarine's predicted position to make an intercept rather than simply following its course.

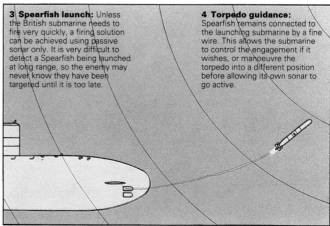

3 Spearfish launch: Unless the British submarine needs to fire very quickly, a firing solution can be achieved using passive sonar only. It is very difficult to detect a Spearfish being launched at long range, so the enemy may never know they have been targeted until it is too late.

4 Torpedo guidance: Spearfish remains connected to the launching submarine by a fine wire. This allows the submarine to control the engagement if it wishes, or manoeuvre the torpedo into a different position before allowing its own sonar to go active.

3 Active

Spearfish's active sonar is mainly used in the last stages of an attack. Modern Soviet submarines, which Spearfish was designed to destroy, are so tough that a proximity explosion which would have destroyed earlier boats would not cause fatal damage. Spearfish's shaped-charge warhead needs to be in contact with the target in order to punch through the tough double hull, and in order to make contact the torpedo has to have very precise guidance, which is provided by powerful bursts of sound which echo off the target back to the weapon. The on-board electronics are sophisticated enough to guide Spearfish to strike the optimum point on a target boat's hull.

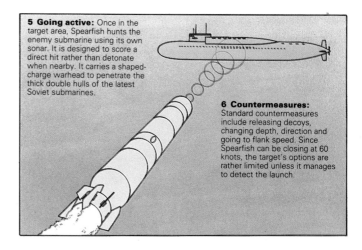

5 Going active: Once in the target area, Spearfish hunts the enemy submarine using its own sonar. It is designed to score a direct hit rather than detonate when nearby. It carries a shaped-charge warhead to penetrate the thick double hulls of the latest Soviet submarines.

6 Countermeasures: Standard countermeasures include releasing decoys, changing depth, direction and going to flank speed. Since Spearfish can be closing at 60 knots, the target's options are rather limited unless it manages to detect the launch.

4 Surface targets

The Royal Navy's fleet submarines are mainly hunters and killers of other submarines, but they are also tasked with destroying surface vessels. As a result, although Spearfish is primarily an anti-submarine weapon, it must also be able to deal with enemy warships. The first torpedoes were designed to smash into the side of a ship, sinking them by blowing a large hole in the side below the waterline. The same effect can be achieved with a smaller explosive charge by detonating it beneath the vessel. The blast and hydrostatic shock create a whiplash effect which breaks the vessel's back, leaving it crippled even if it does not sink. Spearfish is fitted with a proximity fuse designed to detonate the warhead in the optimum position for such an effect.

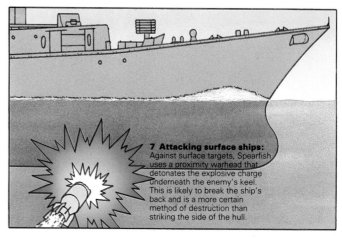

7 Attacking surface ships: Against surface targets, Spearfish uses a proximity warhead that detonates the explosive charge underneath the enemy's keel. This is likely to break the ship's back and is a more certain method of destruction than striking the side of the hull.

An 'Osa' class fast attack craft launches an SS-N-2 'Styx'. Note the exhaust plumes from the rocket booster in the missile body and from the turbojet underneath. Before the development of electronic and other countermeasures, the 'Styx' achieved more than an 80 per cent hit rate.

SS-N-2 'STYX'
ROCKET FROM HELL

On 21 October 1967 the Israeli navy's flagship, the destroyer Eilat, was patrolling off the Egyptian naval base at Port Said. A 1,700-ton destroyer armed with four 4.5-in guns and eight 21-in torpedo tubes, Eilat was formerly the Royal Navy's HMS Zealous, built at Birkenhead from 1942-44. Cruising about 14 nautical miles (25 kilometres) from shore, the destroyer was suddenly struck by three large missiles, each weighing over two tons. She had no warning and, even if she had, Eilat possessed no weapons capable of stopping them. Eilat sank and claimed the dubious distinction of being the first warship to be lost to a ship-launched surface-to-surface missile.

The missiles that sank Eilat were made in the Soviet Union and were of the type known today as the SS-N-2 and given the NATO reporting name 'Styx'. In 1967 very little was known about the 'Styx' outside the world's naval intelligence communities. The famous reference book *Jane's Fighting Ships* had no name for it in the 1966-67 edition, simply noting that Soviet missile patrol boats of the 'Osa' and 'Komar' classes were equipped with a missile 'of 10 to 15 miles' range'. Unfortunately for the crew of Eilat, the

'Styx' turned out to have a range of 25 miles.

The USSR had supplied Egypt with three 'Komar' class missile boats five years earlier. Two had participated in the attack on Eilat, both firing two missiles, so 'Styx' had achieved a 75 per cent hit rate on its first combat engagement.

Turbojet powered

The 'Styx' missile is the size and shape of a small aircraft. With swept-back delta wings and an aircraft tail assembly, it has a turbojet fitted underneath the body. Its overall weight is 2300 kilograms, which includes a 400-kilogram warhead. Its solid propellant rocket booster launches it towards the target at just over 500 knots (925 km/h); the missiles that sank Eilat took less than two minutes to hit their target.

'Styx' is controlled by a radar that has the NATO reporting name 'Square Tie'. The radar detects the target and tracks it while 'Styx' is launched, and the missile's autopilot keeps 'Styx' on a pre-programmed course, flying at an altitude of several hundred metres. As it nears the target, the 'Styx's own radar switches on and the missile dives to attack the biggest radar re-

Below: Weighing more than three times as much as the Exocet of Falklands fame, the 'Styx' is capable of causing catastrophic damage to the largest modern warships.

type boats of the Egyptian and Syrian navies. They all missed. By contrast, the Israelis sank half-a-dozen enemy vessels with their own shorter-ranged Gabriel missiles.

In their way these Arab-Israeli naval actions off Latakia and Baltim were as significant as the first clash of the aircraft carriers at Coral Sea. Neither side saw each other; the battles were fought and won by monitoring enemy radar emissions and firing missiles at anonymous radar contacts. They established the missile as the primary weapon for surface action. Naval guns were now secondary weapons.

Left: The crewmembers on board this 'Osa' class are dwarfed by the massive launcher bins for the four 'Styx' missiles. The turret houses two radar-directed 30-mm cannon.

Specification
SS-N-2a 'Styx'

Type: surface-to-surface anti-ship missile
Dimensions: length 6.3 m; diameter 750 mm; span 2.75 m
Weight: 2300 kg
Warhead: 400-kg HE
Propulsion: solid-propellant rocket booster and turbojet engine sustainer
Performance: speed Mach 0.9; range 40 km
Guidance: command with terminal active radar homing

flection it detects.

The stunning success of the 'Styx' missile spurred the US Navy to develop a similar weapon. Although the US Navy could defend itself with its excellent SAMs – USS *Sterett* shot down an incoming 'Styx' off North Vietnam with a Terrier missile in 1972 – it planned to attack enemy warships with its enormous force of carrier-borne aircraft. The Navy was already developing the AGM-84 Harpoon as an air-launched anti-ship missile and the programme was soon expanded to allow Harpoon to be fired from surface ships. Meanwhile, 'Styx' continued to perform well: in 1971 the Pakistani destroyer *Khaibar* met the same fate as the *Eilat* when she ran into Indian missile boats blockading Karachi. Like the *Eilat*, *Khaibar* was a wartime British destroyer (ex-HMS *Cadiz*) armed with guns and torpedoes. The Indian missile boats remained out of reach of *Khaibar*'s weapons and sank her without difficulty.

Gabriel missile

The Israelis studied 'Styx' intensively. They developed their own anti-ship missile, the Gabriel, and various defensive systems, including electronic countermeasures to jam Styx's homing radar. They manufactured chaff rockets that scattered big clouds of tin foil – clouds that both the missile and the 'Square Tie' radar sets would mistake for a ship.

The Israeli countermeasures proved astonishingly successful during the 1973 Yom Kippur War. Some 54 'Styx' missiles were fired in a series of engagements between Israeli fast attack craft and 'Osa' and 'Komar'

Above: A wooden-hulled 'Komar' class attack craft of the type that sank the Israeli destroyer Eilat *with a salvo of 'Styx' missiles.*

Below: The Eilat *was an ex-British 'Z' class destroyer built during World War II. It had no defence against guided missiles.*

Specification

Standard SM-2ER

Type: long-range naval SAM
Dimensions: length 7.98 m; diameter 34.3 cm (missile), 45.7 cm (booster); span 91.5 cm
Weight: 1510 kg
Performance: speed Mach 3.5; range 150 km; maximum engagement altitude 30000 m
Guidance: inertial with mid-course correction and semi-active radar homing
Warhead: high-explosive fragmentation

America's World War II experience in the Pacific against Japanese kamikaze attacks was a powerful incentive towards the development of effective anti-aircraft systems. Piling as many anti-aircraft guns onto a ship as it could carry was one answer, but that was only partially effective and called for a lot of manpower. Clearly something radically different had to be tried.

'Project Bumblebee'

In the last years of the war, the US Navy initiated a weapons development programme known as 'Project Bumblebee'. An astonishingly ambitious programme, 'Project Bumblebee' was to become one of the most influential missile programmes of all time. Even today, most US Navy ship-launched surface-to-air weapons are descended from the series of missiles developed in the years after World War II.

Above: A Standard surface-to-air missile is fired from the twin-rail launcher of USS Wainwright.

Right: USS St Lo burns after being hit by a Japanese suicide plane off Leyte Gulf. These kamikaze attacks led the US Navy to demand SAMs.

Terrier, Tartar and Talos all entered service in the 1950s and they were the backbone of US Navy air defence until the 1970s. Terrier was a long-range, rocket-powered weapon which became operational in 1956; Tartar was similar but, lacking Terrier's booster motor, was a shorter range weapon; and Talos was a huge, ramjet-powered, long-range missile which saw action off Vietnam and which served from 1958 until the late 1970s.

Development of the Standard missile as a replacement for Terrier and Tartar began in the early 1960s. Very similar in appearance to the missiles it has replaced, Standard is vastly more capable. It entered service from 1968, replacing Terrier and Tartar on existing warships as they came in for refits.

Standard has a cylindrical body with a pointed nose and, as with the preceding missiles, it has in-line, long-chord, narrow-span wings with cruciform trapezoidal control fins at the tail. The missile is semi-active radar homing, originally using a conical scan to home in on the radar transmissions of the launch ship, but in its later versions using a digital computer system to home in on digital monopulse transmissions. Some missiles have been fitted with anti-radar sensors, homing in on enemy radar transmissions.

Vertical launch system

Standard was originally launched from fully automatic single or twin missile launchers, but its most recent application is in the US Navy's latest vertical launch missile systems.

Standard has developed into a family of missiles. Tartar's replacement was the RIM-66A Standard-1MR, which was designed with an engagement range of 50 kilometres and could intercept targets between altitudes of 45 metres and 15000 metres. The fitting of a new dual-thrust motor has increased the range by 45 per cent and the engagement altitude by 25 per cent. The Terrier's replacement was the RIM-67A Standard SM-1ER, basically the same missile but with a booster motor to extend engagement range out to 75 kilometres and altitude up to 25000 metres.

The missile system has undergone considerable change over the years, and the current versions are the RIM-66C Standard SM-2MR and the RIM-67B Standard SM-2ER.

The SM-2MR was developed for use in the US Navy's AEGIS air defence system. With improved electronics and propulsion, the AEGIS system

allows for mid-course guidance which extends the missile's range out to 75 kilometres, double the range of the original Terrier long-range missile.

SM-2ER has also been fielded, but this is not for use in the AEGIS system. Improvements in the missile's booster increases the missile's speed by a full Mach number and it has a range far

Able to pick off aircraft at very high altitudes, Standard can also engage targets as low as 50 metres, so all but surface-skimming missiles can be attacked.

outside the current control limits of the AEGIS system. However, its improved capabilities make it suitable to engage high-performance aircraft and missile targets, and cruisers originally equipped with the extended-range Terrier now use Standard to protect high-value targets such as aircraft carriers.

*USS **T**iconderoga **fires a Standard SM-1MR in 1982 before the cruiser had commissioned. The spectacular launch signature is typical of modern naval surface-to-air missiles.***

Specification
Standard SM-2MR

Type: medium-range naval SAM
Dimensions: length 4.72 m; diameter 34.3 cm; span 91.5 cm
Weight: 708 kg
Performance: speed Mach 2.5+; range 75 km; maximum engagement altitude 20000 m
Guidance: semi-active radar homing
Warhead: high-explosive fragmentation

STINGRAY

Above and below: Stingray is carried by surface ships, submarines and aircraft. It has its own sonar to home in on enemy submarines.

The technological revolution which has taken place over the last four decades has seen a profound change in the waging of war at sea. The lightweight guided torpedo has taken the place of the depth charge as the primary anti-submarine weapon arming surface combatants, whether launched from the ship itself or from one of its embarked helicopters.

With the deployment of the Stingray lightweight ASW torpedo by the Royal Navy in the early 1980s, the era of the 'smart' torpedo began. Stingray is the first British torpedo to be developed entirely by private industry, and it incorporates a number of technical innovations. The weapon can be launched from helicopters, aircraft and surface ships from a wide range of launch speeds and in an equally wide variety of sea states. As a result of its unique guidance system it can be used satisfactorily in both shallow and deep waters, with an equally high single-shot kill probability.

Although deployed aboard several of the Royal Navy's warships in the South Atlantic in 1982, Stingray saw no combat use, and did not in fact achieve full-scale service with the Royal Navy and the Royal Air Force until 1983. Since then, it has been sold to Egypt, Norway and Thailand.

In terms of general performance, Stingray is similar to the American Mk 46, which it has replaced in British service, and is in widespread use around the world. However, the British weapon has a deeper diving capability. Where Stingray scores most highly in comparison with other such weapons is in its greatly advanced guidance system.

The torpedo's on-board digital computer coupled with a multi-mode, multi-beam, active/passive sonar effectively make Stingray a 'smart' weapon, capable of making repeated attacks on manoeuvring targets. Propulsion is provided by an electrically-driven pump-jet with a sea water-activated battery, which ensures that there is no speed loss with increasing depth and water density.

Anti-submarine weapon

Stingray was intended, when designed in the late 1970s, to deal with the latest Soviet submarines. These were fast and deep-diving. In addition, their double-hull construction made them very tough, possibly tough enough to survive the effects of the explosion of a conventional torpedo warhead. As a result, Stingray's warhead is of a directed-energy, shaped-charge type, designed to punch through both inner and outer hulls of a Soviet submarine.

The end of the Cold War might have decreased the threat from former Soviet forces, and so reduced the need for high-tech weaponry like Stingray. Appearances can be deceiving, however. At the last count there were more than 200 submarines in service around the world outside the major navies. Some of the boats are old, but many are new, high-tech vessels. A high proportion of these are operated by regimes whose stability and friendliness to the West are a matter of some debate. As long as there is a threat, there will be a need for weapons like Stingray.

Specification
Stingray

Dimensions: length 2.6 m; diameter 32.4 cm
Weight: 265 kg
Performance: speed 83 km/h; diving depth 800 m; range 11 km
Warhead: 40-kg high-explosive shaped-charge

Engagement sequence

1 Carriage

On surface ships Stingray is launched from triple torpedo tubes. These are quick-reaction launchers, tied directly into the launch vessel's Action Information Organisation systems. Stingray will more often be deployed from the vessel's anti-submarine warfare (ASW) helicopter. Two Stingrays are a typical load for the Westland Lynx carried on the Royal Navy's smaller ships, while the larger Westland Sea King and the EH.101 Merlin helicopters can each carry four torpedoes. The Royal Air Force has also adopted Stingray for use by its Nimrod maritime patrol aircraft: each Nimrod MR.Mk 2 aircraft can carry up to nine torpedoes in its weapon bay.

2 Tracking

Under normal circumstances at sea the computerised shipboard tracker and predictor system receives information from the main action information system, which collates data from shipborne and heli-borne sonars. The target's current position is then calculated. Repeated measurements give the target's course and speed, enabling a course prediction to be made. Data is continually fed into the torpedo's guidance system, and once optimum launch position is reached, the torpedo is ejected from its tube or dropped by helicopter, roughly in the direction of the target. It is retarded in flight by a small drogue parachute, which ensures that the weapon adopts the correct nose-first attitude on entering the water.

3 Guidance

Once the weapon is submerged, the parachute is discarded and the weapon makes for the target's predicted location, where it begins to travel in a search pattern while listening for the target. The on-board computer and multi-beam active/passive sonar enable Stingray to make its own tactical decisions during an engagement, by anticipating the target's changes of course, speed and depth through analysis of the information it receives via its sensors. The computer is also intelligent enough to ignore countermeasures such as towed noise-makers, acoustic jammers or mobile decoys that may be deployed by the target boat.

4 Homing

Stingray is tenacious: once locked onto a validated submarine target it is almost impossible to shake off unless the submarine is fast enough and has enough warning to get out of range before being hit. However, Stingray's pump-jet propulsion makes it very difficult to hear. The computer guides the torpedo to the optimum point on the target, where the shaped-charge warhead is exploded by a contact fuse so that all of the power of the explosion is directed into penetrating the pressure hull. In deep water, the immense water pressure will do the rest, while in shallow water the blast should do enough damage to force the target boat fatally deeper, or to drive it damaged and vulnerable to the surface.

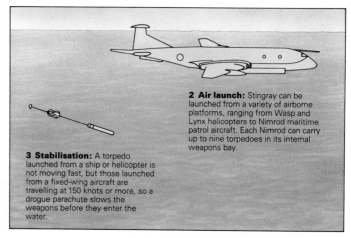

1 Ship launch: Stingray torpedoes are carried and launched from surface ships in a triple mount derived from the US Navy's Mk 32 torpedo launch unit. The ship's Torpedo Weapon System-2 is a quick-reaction unit which is integrated into the carrier vessel's Action Information Organisation system.

2 Air launch: Stingray can be launched from a variety of airborne platforms, ranging from Wasp and Lynx helicopters to Nimrod maritime patrol aircraft. Each Nimrod can carry up to nine torpedoes in its internal weapons bay.

3 Stabilisation: A torpedo launched from a ship or helicopter is not moving fast, but those launched from a fixed-wing aircraft are travelling at 150 knots or more, so a drogue parachute slows the weapons before they enter the water.

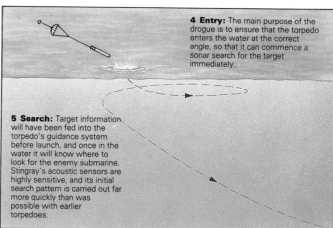

4 Entry: The main purpose of the drogue is to ensure that the torpedo enters the water at the correct angle, so that it can commence a sonar search for the target immediately.

5 Search: Target information will have been fed into the torpedo's guidance system before launch, and once in the water it will know where to look for the enemy submarine. Stingray's acoustic sensors are highly sensitive, and its initial search pattern is carried out far more quickly than was possible with earlier torpedoes.

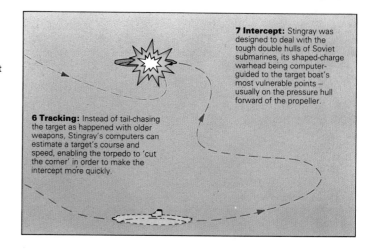

7 Intercept: Stingray was designed to deal with the tough double hulls of Soviet submarines, its shaped-charge warhead being computer-guided to the target boat's most vulnerable points – usually on the pressure hull forward of the propeller.

6 Tracking: Instead of tail-chasing the target as happened with older weapons, Stingray's computers can estimate a target's course and speed, enabling the torpedo to 'cut the corner' in order to make the intercept more quickly.

The modern supercarrier is the largest, most expensive and most powerful conventional warship ever built. With a crew of 6,300 and an air wing of 80 aircraft, the carrier is designed to face any land, sea or air threat.

SUPERCARRIER

Supercarriers like the USS Nimitz are the most powerful and flexible warships ever built. They can respond to any situation, from port visits and 'show the flag' cruises, through disaster relief and civilian evacuation, right up to fleet battles and nuclear war. That power and flexibility is provided by the 80 aircraft of the carrier's Air Wing.

The USS Dwight D. Eisenhower (CVN-69) and the USS Nimitz (CVN-70) are seen during an Indian Ocean battle group changeover some time in the early 1980s. The escort is the nuclear-powered guided missile cruiser USS South Carolina.

But USS *Nimitz* and her sisters are much more than floating aerodromes; 3,000 officers and men are required to keep the aircraft flying and fighting. Pilots may be the 'Top Guns' of the business, but without the maintenance men, the ordnance handlers, the flight deck crew and all the other support personnel, they would be about as much use in the air as a penguin.

A carrier's flight deck is big: over 1,000 feet long and 250 feet wide, it is as large as three football pitches. But when you are operating 80 aircraft it can get pretty crowded. And working on that deck is hot, dirty, and poten-tially deadly. 'Deck Apes' come in a variety of colours, wearing shirts which instantly indicate their function. Green Shirts are the maintenance crew, who do a variety of jobs including the highly dangerous ones of hooking aircraft onto the catapult at launch and checking whether the tail hook has safely caught on recovery.

Yellow Shirts direct traffic and are responsible for the delicate task of moving 80 or more aircraft around on limited flight deck space. It is the yellow-shirted catapult officer who gives the final clearance for an aircraft to launch. Brown Shirts are the aircraft maintenance crew, looking after their charges like a mother looks after her babies. Blue Shirts are the shunters and movers, the tractor drivers and the like.

No matter what the colour of the

Aboard the supercarrier

The deck of a supercarrier is a fifth of a mile long, and covers more space than three football pitches. Vast though it is, it is still very small when you consider that it has to serve as an airport for more than 80 large combat aircraft. Thanks to careful design and organisation, however, it is possible to launch planes, land them and park them, all at the same time.

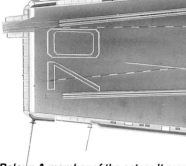

Below: A member of the catapult crew uses hand signals as an A-6 is hooked up for launch. This is the only way to communicate in an environment dominated by screaming jet engines.

Catapults
The carrier's four catapults are located in the bow and at the front of the angled deck. Powered by steam drawn from the ship's huge boilers, they are each capable of accelerating a 30-ton aircraft to flying speed in about 1½ seconds.

Angled deck
The simple device of swivelling the landing on deck a few degrees to the left means that any planes landing on a modern carrier can accelerate away if the landing is wrong, without the danger of running into stationary planes on deck. Its additional catapults also double the number of aircraft that can be launched at any one time.

shirt, one thing is constant: a carrier's deck is dangerous, and it takes teamwork to ensure accidents do not happen.

Movement and maintenance of aircraft are important, but a warplane needs more than these in order to fly and fight. The Red Shirts are 'Ordies', or ordnance handlers. They look after the aircraft's weapons, taking them from the magazines and loading them onto the aircraft. They deal with missiles and cannon ammunition for the fighters, tons of bombs and air-to-surface missiles for the attack birds, and torpedoes, depth charges and the like for the carrier's anti-submarine aircraft and helicopters.

The Purple Shirts are another team in a dangerous trade, and are responsible for fuelling the aircraft. Both Red Shirts and Purple Shirts have to be highly safety-conscious. Fire is a real danger aboard a carrier, packed as it is with highly explosive stores, so fuelling aircraft is carried out with as much care as possible. On the rare occasions that there are accidents, it is usually under combat conditions. It was accidental firing of weapons which caused the disastrous fires aboard the USS *Oriskany*, *Forrestal* and *Enterprise* during the Vietnam War.

While the aircraft and the personnel of the air group are vital to the vessel's fighting ability, it should be remembered that an aircraft carrier is first and foremost a very large ship, and it requires an equally large crew to make it run. The 3,300 men of the ship's complement handle everything from manning the nuclear reactors and steam turbines that power the ship, to providing all the services that a community of 6,300 requires.

A carrier's hangar cannot shelter the entire air wing. Some aircraft remain on deck and need to be secured. These brown-shirted plane captains are standing by with tie-down chains.

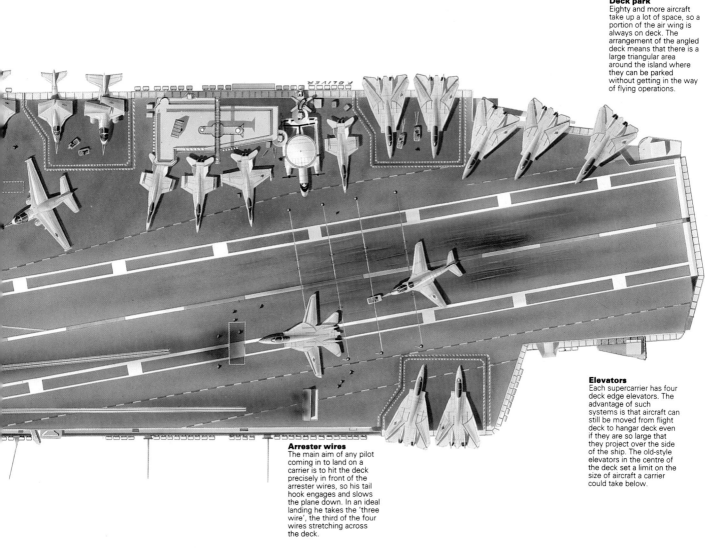

Deck park
Eighty and more aircraft take up a lot of space, so a portion of the air wing is always on deck. The arrangement of the angled deck means that there is a large triangular area around the island where they can be parked without getting in the way of flying operations.

Elevators
Each supercarrier has four deck edge elevators. The advantage of such systems is that aircraft can still be moved from flight deck to hangar deck even if they are so large that they project over the side of the ship. The old-style elevators in the centre of the deck set a limit on the size of aircraft a carrier could take below.

Arrester wires
The main aim of any pilot coming in to land on a carrier is to hit the deck precisely in front of the arrester wires, so his tail hook engages and slows the plane down. In an ideal landing he takes the 'three wire', the third of the four wires stretching across the deck.

An aircraft carrier is just that: a ship which carries, launches and recovers aircraft. All the space which can be is dedicated to the operation of its air wing. But it is still a warship. The seamen who keep the ship running work shifts around the clock, as they would on any other fighting vessel. All the usual tasks have to be performed, from steering and navigation, to hull-painting and anchor maintenance. The only thing to break the flight deck is the carrier's island, and it is from here that its operations are controlled.

The sheer size of a supercarrier is awesome. At 90,000 tons, it is the largest warship ever built. It is also one of the fastest ships afloat, and can force its 90,000 tons through the water at nearly 65 kilometres per hour (40 mph). Obviously, something pretty potent must be happening down in the engine rooms. All carriers are steam-powered, although that steam can be generated by a nuclear reactor

The Landing Signal Officer is an experienced pilot who is in ultimate control of all aircraft-trapping aboard a carrier. He also grades every landing, so that the Air Wing Commander has a continuous measure of pilot performance.

or by conventional boilers. The steam drives four huge turbines, turning four propellers, and the entire system generates over 280,000 shaft horsepower.

Looking after these mighty machines is another highly-skilled trade, one that calls for men to work under conditions of high temperature and almost deafening noise. Without those men, however, steam would not

be generated and the ship would not move. Without steam, there would be no fresh drinking water, no catapult launches, no heating or air-conditioning, and no electricity to light the vessel or to cook the 20,000 meals a day that are necessary to feed a crew of 6,300 hungry men. It is hardly a glamorous job, but working in the engineering spaces is vital.

shop. The ship even has its own television station. Such comforts might be thought overly luxurious on a fighting ship, but they are absolutely vital in a vessel which can be on deployment, far from home and loved ones, for six months at a time.

On-board community

Administration of such a complex system is a major task in itself. The supercarrier community is run like a major corporation, with the Captain as chairman of the board and the Executive Officer (known as 'XO') as managing director. The ship is split into departments, like the operating divisions of a company. The departments are Air, Air Wing, Aircraft Maintenance, Operations, Navigation, Communications, Engineering, Safety, Administration, Supply, Training, Medical and Dental.

There is often another level of command, however. Operating at the heart of a carrier battle group, the carrier will usually be home to a Flag Officer and about 70 of his staff. The Admiral provides the long-range planning, deciding what he wants the carrier force to do, with his staff developing the plans which will enable the battle group to do it. However, the Captain remains in sole command of the vessel itself, and he makes the decisions on how the ship is used to complete any task assigned by the battle group commander.

At any one time, a carrier is a weapons platform, an oil tanker, a town of 6,000 souls, a factory and an airport. These characteristics and the skills of the men who make up the crew go together to make it the largest, most powerful, most flexible warship ever to sail the seas.

Big though it is, a carrier is not the most spacious of vessels to live on. Eighty aircraft take up a lot of room, to say nothing of the equipment to keep them running and fighting, the massive engines which power the carrier, millions of gallons of aviation fuel, and thousands of tons of stores and supplies. There is little privacy for any but the most senior officers of the 6,300 men who make up the carrier community, yet every effort is made to make life bearable. Just feeding such a crew is a major task, but food is plentiful and varied. Even so, the fact that the ship's bakery prepares over 6,000 hamburger and hot dog rolls every day indicates that fast food is as popular aboard ship as it is back home. Some canteens are open round the clock to cater for sea-going watch routines.

Other on-board facilities include hospitals, laundries, shops, movie shows, and that most important of military establishments, the barber's

Above: The carrier's superstructure carries all of the myriad radar, electronic warfare and communication antennas necessary to fight a capital ship in modern battle.

Below: The product of all the carrier's sensors goes to the combat information centre deep in the carrier's hull, from where the vessel is controlled in battle.

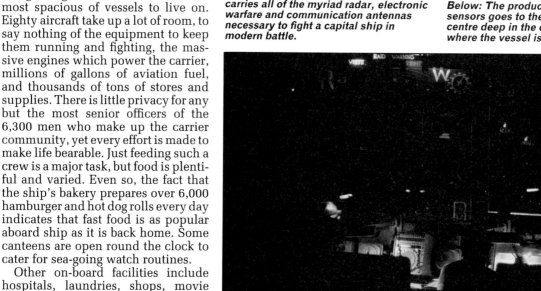

TRIDENT
Destroyer of Worlds

The purpose of the Trident programme was to increase the range of American sub-launched missiles, thus allowing the use of larger and more remote patrol areas. Trident I was tested in 1977, becoming operational two years later aboard converted SSBNs of the 'Benjamin Franklin' and the 'Lafayette' classes. The system uses the three-stage solid propellant UGM-96A C-4 missile.

Increased range was considered more important than accuracy, which was about the same as that of the preceding 'Poseidon' class, 50 per cent of missiles are likely to strike within 500 metres of their targets. But that accuracy was over a range increased from about 5000 kilometres to nearly 7000 kilometres.

In the mid-1970s, a programme to improve Trident's accuracy was set in motion. The Trident II utilises the D-5 missile, which is considerably larger than the preceding C-4. It has a greater range, and much improved accuracy. That accuracy is such that for the first time the US Navy has the capability to strike at hardened targets such as enemy missile silos and command bunkers, previously only possible with land-based missiles.

The programme has also increased the number of warheads that the missile can carry, from the eight 100-kiloton weapons of the C-4 to a maximum of 12 475-kiloton weapons.

Cold comfort

The D-5 can only be fired from 'Ohio' class missile boats. The original intention was to build at least 20 of these huge submarines, ordering over 800 missiles to support them. With the end of the Cold War, it is likely that the 18 boats already in service, under construction or funded by Congress, are going to be all there are.

Even with reduced numbers, Trident remains an awesome system. Eighteen boats can carry 432 missiles, which in turn can carry more than 5,000 warheads, each of which is 15 times more powerful than the bomb which destroyed Hiroshima in 1945.

A single 'Ohio' class ballistic missile submarine has more explosive power than all of the bombs dropped in World War II, Korea and Vietnam combined, while a single Trident II warhead is equivalent to four times the entire weight of explosive dropped on Kuwait and Iraq during the Gulf War.

Trident D-5 is the ultimate deterrent: launched from submarines which are almost impossible to locate before they fire, it has a range of 12000 km and is accurate enough to destroy hardened targets.

Specification
D-5 Trident II

Type: submarine-launched ballistic missile
Dimensions: length 13.42 m; diameter 2.11 m
Launch weight: 59090 kg
Propulsion: three-stage solid propellant rocket
Performance: maximum range 12000 km
Guidance: stellar/inertial
CEP: 100 m
Warhead: Mk 5 re-entry system with between eight and 12 re-entry vehicles, each with 300-475-kiloton warheads

Engagement sequence

1 Launch order

The trouble with communicating with submarines is that they are under water, and water is almost totally opaque to normal radio waves. However, ELF, or extremely low frequency, waves are not stopped by water, but they don't carry much information. It can take several minutes just to get a three-letter code group through. However, that is enough for a few simple messages, which is adequate to instruct the boat to come to periscope depth. Once there, a needle-thin ESM mast is raised, which is equipped with sensitive detectors. If no hostile radar transmissions are detected, the periscope is raised to check that the surface is clear, and then the radio mast will be raised. Now the boat can receive satellite messages, which go through a rigorous process of authentication.

2 Launch and boost

Once the captain and his senior officers agree that the order to launch has indeed come from the proper authorities, which in most cases means directly from the President, they will prepare to launch. A number of keys must be turned simultaneously to allow the missiles to be armed and launched, which is a precaution to prevent a madman from starting World War III. The outer doors are opened, the missile tube is flooded, and the 24,000-ton hull of the boat shudders as the missile is blasted free from its tube by compressed air. The first-stage rocket ignites as the missile breaks the surface, and a spike deploys from the nose cone, which improves the Trident's aerodynamic performance. The second and third stages boost the missile into its sub-orbital path, well beyond the atmosphere.

3 Post-boost

Once the three stages of the rocket motor are burned out, less than three minutes after launch, the missile enters the post-boost phase. Previous missiles were fitted with an inertial navigation system. Interfacing with the launch boat's navigation system, the guidance system knows where it has been launched from and to where it has to go, and calculates its flight path accordingly. No matter how good an inertial system is, however, errors will creep into the calculation, and inaccuracies will result. Trident's accuracy arises from the fact that it does not rely on inertial navigation alone. The Mk 5 guidance system contains a star sensor, which it uses to take navigational fixes from stars while in flight, and the missile will correct its flight path using that data.

4 Re-entry

The post-boost control system has thrusters which allow the final stage of the missile to make corrections for errors in launch position data. It also manoeuvres to deploy the re-entry bodies, each of which carries a warhead. The warheads are unpowered; where they are released from the post-boost vehicle determines where they will land. It was originally planned to fit Trident with the Mk 4 Evader MARVs, or manoeuvring re-entry vehicles. MARVs would be capable both of in-flight manoeuvre to avoid enemy anti-missile defences, and of terminal in-atmosphere guidance to give a very low CEP, or circular error probable, of about 20 metres. CEP is a statistical measure of accuracy, and is the radius of a circle centred on the target within which 50 per cent of warheads can be expected to hit.

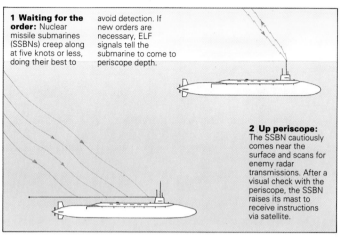

1 Waiting for the order: Nuclear missile submarines (SSBNs) creep along at five knots or less, doing their best to avoid detection. If new orders are necessary, ELF signals tell the submarine to come to periscope depth.

2 Up periscope: The SSBN cautiously comes near the surface and scans for enemy radar transmissions. After a visual check with the periscope, the SSBN raises its mast to receive instructions via satellite.

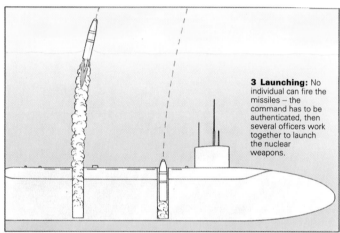

3 Launching: No individual can fire the missiles – the command has to be authenticated, then several officers work together to launch the nuclear weapons.

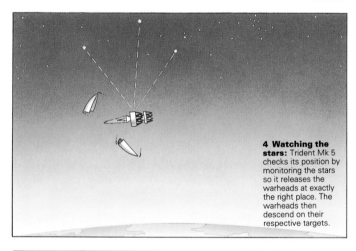

4 Watching the stars: Trident Mk 5 checks its position by monitoring the stars so it releases the warheads at exactly the right place. The warheads then descend on their respective targets.

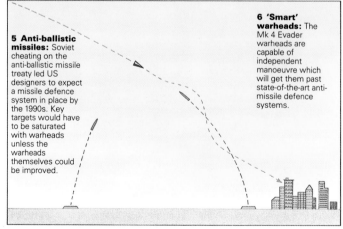

5 Anti-ballistic missiles: Soviet cheating on the anti-ballistic missile treaty led US designers to expect a missile defence system in place by the 1990s. Key targets would have to be saturated with warheads unless the warheads themselves could be improved.

6 'Smart' warheads: The Mk 4 Evader warheads are capable of independent manoeuvre which will get them past state-of-the-art anti-missile defence systems.

TYPE 21 FRIGATES

HMS Arrow provided naval gunfire support to the British ground forces during the final battles outside Port Stanley. Arrow helped fight the fires aboard HMS Sheffield after the Exocet attack on 4 May 1982 and took off the survivors when the destroyer had to be abandoned.

Since the first Type 21 frigate, HMS Amazon commissioned in 1974, the class has proved very popular with its crews. The tasks with which they have been involved have included the peacetime missions of training, 'showing the flag' abroad, shadowing Soviet warships and undertaking security patrols in such places as the Persian Gulf to protect British shipping from 'accidental' attack by Iran and Iraq during the war in the Gulf. However, the class' first claim to fame actually occurred in 1977 when the *Amazon*, on a deployment to the Far East, suffered a bad internal fire which highlighted the risk of building modern warships with aluminium superstructures to save top-weight.

It was not until the 1982 Falklands war that the Type 21s finally managed to prove their worth. HMS *Antelope*, HMS *Alacrity* and HMS *Arrow* sailed in the early part of April as part of the Task Force's escort and to convoy high-value surface ships such as the troopship *Canberra*. Once the Total Exclusion Zone had been entered, a surface bombardment group, comprising the destroyer HMS *Glamorgan* with the *Arrow* and the *Alacrity*, was ordered to bombard Argentine positions around Port Stanley on the afternoon of 1 May, following the Avro Vulcan and BAe Sea Harrier air strikes of the morning.

HMS Arrow *hit*

During the withdrawal the *Arrow*, which had just become the first ship to fire at the islands, also became the first Royal Navy vessel to suffer damage in the conflict when the group was attacked by a number of IAI Daggers using bombs and cannon fire: a number of 30-mm rounds hit her superstructure, injuring one rating.

Then, on 4 May, the destroyer HMS *Sheffield* was hit by an air-launched AM.39 Exocet missile. In a brilliant display of seamanship the captain of the *Arrow*, Commander P. J. Bootherstone, brought his vessel alongside the burning destroyer to initially help fight the fires raging uncontrollably below decks and then, when it had been realised that all was lost, to take off the survivors. During the rescue operation that accounted for 236 members of a total ship's company of around 300, HMS *Yarmouth* fired her Limbo ASW mortar at a suspected submarine contact and reported seeing a torpedo track near to the vessels. This attack has not been specifically claimed by the Argentines, who will only admit that their submarine *San Luis* made three torpedo attacks against Task Force ships during the war, two against surface ships and one against a possible submarine.

The *Arrow* then went on eventually to join the San Carlos gun line after the initial British landings and to provide

Above: With a third of her company killed or wounded, HMS Ardent stayed in action until her steering failed and the fire was out of control.

Below: Seen here with HMS Plymouth, HMS Avenger was attacked by four A-4s on 30 May; two were shot down and the bombs missed. The Argentines claimed to have sunk Invincible.

the vital naval gunfire support to the 2nd Parachute battalion's epic battle at Goose Green. The *Alacrity* in the meantime was employed in operations to prepare for the main 21 May assault at San Carlos. She landed the Special Boat Squadron reconnaissance team responsible for looking at potential landing beaches, and on the night of 10 May sailed all the way through Falkland Sound between the two main islands on a mission designed to both harass the enemy and to locate shore defences or minefields that might endanger the operation. During the sweep she discovered an unidentified ship which failed to answer challenges. This vessel was then engaged with the *Alacrity's* single 4.5-in gun on rapid fire. The scoring of several hits resulted in the ship exploding in a sheet of flame and rapidly sinking. Subsequently it was found that she had been the Argentine naval transport *Islas de Los Estados* carrying a cargo of aviation fuel and other supplies for the garrison.

Escort duty

It was also *Alacrity's* unfortunate duty to be one of the ships escorting the *Atlantic Conveyor* on 25 May, when she was hit by an air-launched AM.39 Exocet after the *Ambuscade* had detected the incoming raid on her Type 992Q radar. Like *Arrow* previously, the *Alacrity* moved alongside

HMS *Ambuscade*

HMS *Ambuscade* commissioned in 1975 and served in the Falklands campaign. She provided naval gunfire support for the Parachute Regiment's attack on Wireless Ridge and conducted several bombardments of Argentine defences around Port Stanley.

Sea Cat missiles
The Type 21 frigates all have a quadruple Sea Cat SAM launcher mounted above the helicopter hanger. Two or four 20-mm anti-aircraft guns are also carried and the forward 4.5-in gun can be used against aircraft as well.

the victim of a bomb attack. Here two bomb disposal men were flown aboard and the crew were brought on deck and moved aft as the disposal men started work on the first bomb, which had come to rest in the vicinity of the engine room. During the defusing process the bomb exploded, killing one officer instantly, injuring the other and blowing a huge hole in the ship's side from the waterline to the funnel.

The explosion also caused a fire,

to fight the fires and then rescue survivors as the crew abandoned ship.

The third ship of the original three Type 21s that went to the Falklands was the *Antelope*, and she was not so lucky. Escorting the amphibious force into San Carlos Water on D-day, she survived the numerous air attacks of that day only to be singled out by a series of McDonnell Douglas A-4 Skyhawk strikes on 23 May when she was positioned in an air-defence slot at the entrance to San Carlos Water. The last two attacks resulted in two bomb hits by single 1,000-lb bombs, the first aft above her engine room and the second forward in the vicinity of the petty officers' mess, killing a steward and injuring two other ratings.

Luckily the bombs did not explode and the *Antelope* was able to move away up San Carlos Water to a position close to the damaged 'Leander' class frigate HMS *Argonaut*, herself

Exocet missiles
Four MM.38 Exocet missiles are carried in single launchers forward of the bridge. They have a range of 42 kilometres and carry a 165-kilogram warhead.

4.5-in gun
Frequently used to support British ground forces in the Falklands, the 4.5-in gun has a range of 22 kilometres. HMS *Alacrity* sank the Argentine transport *Islas de Los Estados* with her 4.5-in gun in the only surface action of the war.

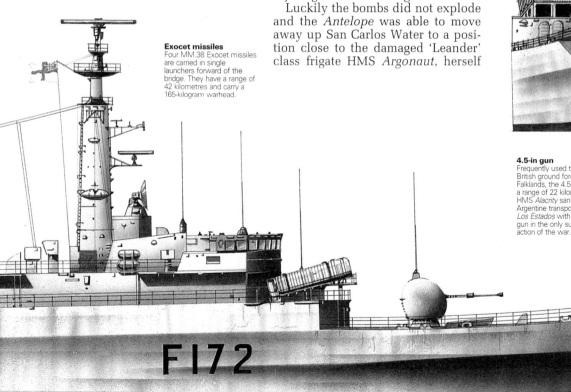

which rapidly spread within the depths of the ship. After fighting the blaze for approximately one hour Commander Nick Tobin, the captain, ordered the ship to be abandoned as the fire was raging out of control. Just after the evacuation had been completed to the various craft around *Antelope*, a series of explosions occurred as the flames reached the ordnance magazines. This caused more fires which gutted the ship, and on the following day she broke in two to form a last defiant V shape before the pieces finally slipped beneath the sea.

Two days before the *Antelope* was hit, the *Ardent* became the second British naval vessel to be sunk by Argentine action. On 21 May she took up an air defence station after bombarding the area of Goose Green. Attacked by Skyhawks late in the afternoon, she was hit near the stern by three 500-lb bombs: these disabled the Sea Cat SAM system and caused a fire which had just been contained by the damage control teams when more Skyhawks attacked. These hit her with a succession of bombs that killed 22 men and wounded 37 others. They also caused a fire which rapidly spread to the whole of the ship's after end.

The damage sustained in the attacks was found to be too much for the repair parties and the order to anchor and abandon ship was given by the captain, Commander Alan West. Most of the *Ardent*'s survivors were taken off the forecastle by the *Yarmouth*, the rest being helicopter-winched from

Above: HMS Antelope's Sea Cat and torpedo magazines exploded after gale force winds fanned the fire resulting from a failed attempt to defuse a bomb.

Below: Men and supplies continue to be shuttled to the beach head at San Carlos while the wreckage of HMS Antelope slowly sinks.

the sea in their 'once-only' survival suits. The derelict hulk, heeling slightly to starboard with the hangar and stern totally wrecked and ablaze, sank soon after this.

Gunfire support

Of the other class members involved in the war, HMS *Avenger* and HMS *Active* were used in the naval gunfire support role during the final battles around Port Stanley. The *Avenger* was also involved in the 30 May AM.39 Exocet attack on the carrier force. During the attack she claimed one Skyhawk downed by fire from her 4.5-in gun and an Exocet missile destroyed by another shot from

the weapon. However, the Skyhawk was credited to a Sea Dart shot from HMS *Exeter* together with a second aircraft, while of the fate of the missile the only thing known for certain is that it did not hit a Task Force ship.

The only vessel of the class not to serve in the Falklands was the *Amazon*, which was on deployment in the Far East. Although suffering two losses and the discovery of hull fracturing as a result of local sea conditions, the Type 21s proved themselves excellent general-purpose frigates despite being somewhat underarmed, and they have all undergone a strengthening programme since returning home.

US BALLISTIC MISSILE SUBMARINES

THE NUCLEAR SHIELD

Since the early 1960s ballistic missile submarines have been a vital part of NATO's nuclear shield. While land-based nuclear missiles and nuclear-capable bombers could be knocked out in a surprise attack, the submarines remained hidden. Lurking in the vast reaches of the Atlantic and Pacific oceans, they could retaliate with enough missiles to obliterate the USSR. Now that the Soviet Union has disintegrated their numbers will decline, but their importance will not. Ballistic missile submarines remain the ultimate military deterrent.

The current US Navy ballistic missile submarine (SSBN) fleet consists of 12 'Ohio' class submarines and 22 of the 'Benjamin Franklin' and 'Lafayette' classes. The massive 'Ohios' are as long as the famous Soviet 'Typhoon' class, although they lack the beam of the double-hulled Soviet vessels. Displacing 16,700 tons, they carry 24 Trident missiles which have eight 100-kiloton warheads each

Right: A Polaris A3 blasts out of the sea during a 1964 test. Undetectable and thus immune to a first strike, the US submarine fleet is the ultimate deterrent against nuclear attack.

Above: Each 'Ohio' class submarine carries 192 warheads, each five times more powerful than the bomb that devastated the city of Hiroshima in 1945.

and have a range of over 7400 kilometres. The phenomenal range of the missiles allows the 'Ohios' to operate far from the USSR and beyond the reach of most Soviet naval units. Another six 'Ohios' are under construction, but it seems unlikely that more will be ordered as the Soviet threat has so diminished.

Ageing submarines

The 'Benjamin Franklin' and earlier 'Lafayette' class SSBNs are all about 30 years old, built as part of the USA's ambitious naval building programme that saw 41 Polaris missile submarines completed between 1959 and 1964. Polaris had a range of under 3000 kilometres and was replaced during the 1970s by Poseidon, which could reach targets up to 5000 kilometres away. The 'Benjamin Franklins' and 'Lafayettes' all served with the US Atlantic Fleet, some operating from Scotland. The US Navy based a submarine tender at Holy Loch on the River Clyde, which enabled its SSBNs

Missile tubes
The *Ohio* has 24 missile tubes, compared with the 16 fitted to earlier US missile submarines. Because of the larger number of missiles, fewer 'Ohios' are required to maintain the seaborne element of the USA's nuclear deterrent.

Engine room
The *Ohio* is driven by a pair of turbines mounted on 'rafts' that isolate them from the hull. This reduces the noise made by the submarine as it manoeuvres underwater.

Above: An officer takes a smoke break outside USS Lafayette's sick bay – in the centre of the missile compartment.

USS *Ohio* began her sea trials. After that the 'Ohios' were commissioned at an average rate of one per year.

In 1963 the British government announced its intention to order four or five 'Resolution' class SSBNs to take over from the RAF's V-bomber force as Britain's nuclear deterrent. The option on the fifth boat was cancelled in 1965: four SSBNs is the minimum necessary to guarantee that one is on station at all times. In February 1968 HMS *Resolution* made Britain's first Polaris missile launch at the US Navy's test facilities in Florida. Although constructed in Britain, the 'Resolutions' are very similar to the American 'Lafayette' class. Even though their missiles have been successively upgraded, these ageing boats could not have continued in service to the end of the century. HMS

to operate in north European waters where their missiles could reach deep into the USSR. Several nuclear submarines were also tasked with the direct support of NATO ground forces in Germany.

The very urgency of the original American SSBN programme has led to block obsolescence. Even if the Soviet Union had survived into the 21st century, the American submarine fleet would have shrunk quickly in the 1990s. Although the contract for the first 'Ohio' class boat was issued in 1974, all manner of problems delayed the work, and it was not until 1981 that

Crewmen eat in the dining hall of USS Ohio during pre-commissioning duty in November 1981. There are 155 men aboard this massive warship.

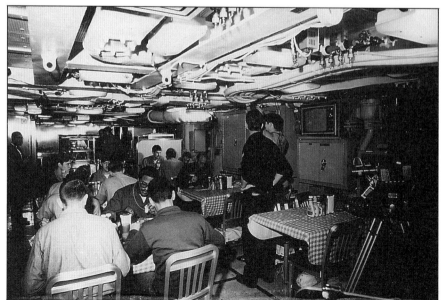

Fin
Following established US Navy practice, the *Ohio* is fitted with fin-mounted diving planes. These are more efficient at low speeds than bow planes, but they limit under-ice operations.

Inside USS *Ohio*

Launched in 1979, the USS *Ohio* was the first of a new class of American missile submarines, more heavily armed than any warship in history. At 18,700 tons it is larger than any other submarine except the monstrous Soviet 'Typhoon' class. By the end of the 20th century the US Navy will have 18 of these boats in service.

Vanguard is the first of four replacements – 16-tube SSBNs twice the size of the 'Resolutions' and equipped with the latest noise-reduction gear to make them harder to detect.

France left NATO's military structure in the 1960s and closed down the US air bases on her soil. Nevertheless, the French navy managed to build five SSBNs beginning with *Redoutable* laid down in 1964 and commissioned in 1971. The *Redoutable* served for 20

years before her withdrawal in 1991 and her four sister-ships are expected to remain operational for another 15 years. A one-off improved submarine 'Inflexible' entered service in 1985 and two new 'Triomphant' class boats are scheduled for launch in 1993 and 1995, respectively.

The French SSBNs are armed with

US sailors train with emergency breathing apparatus aboard USS Daniel Boone (SSBN-629).

nuclear missiles designed and built in France. Each 'Redoutable' carried 16 M20 missiles until 1985; these had a range of just over 3000 kilometres and carried a monstrous 1.2-megaton warhead. The M20s were replaced by the longer-ranged M4, which can travel 4000 kilometres and carries six 150-kiloton MRVs (multiple re-entry vehicles).

The US, British and French ballistic missile submarines are all operated by two crews which take it in turns man-

The crewmens' sleeping quarters may look spartan, but they are much better than on earlier submarines.

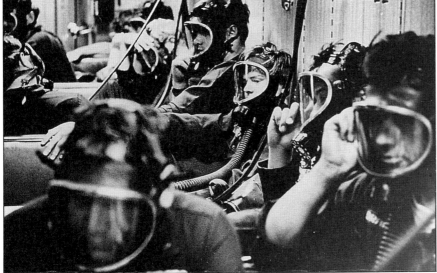

ning the boat. This maximises the time the submarines can spend at sea. All three navies take great care to conceal the operational areas of their SSBNs, which spend much of their patrol time creeping at slow speed, listening for the approach of Soviet attack submarines. Remaining submerged in the deep waters of the North Atlantic, the SSBNs trail long towed-array sonar lines behind them. These are capable of detecting sonar contacts out to 100 nautical miles in good conditions. The submarines make no radio transmissions that might betray their location. Their only contact with the outside world takes place via ELF (extremely low frequency) antennas. These are floated to the surface on a buoy to receive operational orders and sometimes brief family messages.

Soviet spying

The Soviets used to devote a good deal of effort to monitoring the SSBNs. The intelligence officers of GRU (Soviet military intelligence) rented houses on the River Clyde to observe the arrival and departure of submarines. Once at sea, Western SSBNs were the target of every reconnaissance system of the Soviet navy. The USSR's military leaders assiduously

planned to fight and win a nuclear war, and the better picture they had of NATO SSBN deployment, the better their chances of sinking the submarines before they could launch their missiles.

As early as the mid-1980s some defence experts were claiming that the Soviets were developing satellite-based radars that could detect submerged submarines down to several

USS Ethan Allen fired the first test missile with a live nuclear warhead in 1962. It detonated in the Christmas Island test area.

hundred metres. In public, at least, the US Navy dismissed this frightening possibility, but recent tests with US submarines and the Russian Almaz satellites have led the Senate Armed Services Committee to initiate emergency measures.

While NATO, France and the USSR have operated SSBNs for 30 years, the fifth nation officially credited with nuclear weapons has only recently commissioned one. China built a copy of the Soviet diesel-electric 'Golf' class missile submarine in 1964, but China's first nuclear-powered ballistic missile submarine was not laid down until 1981 and only commissioned in 1987. A second hull was reportedly launched in 1981, but its current status is unclear. The 'cultural revolution' inflicted grave damage on China's scientific community, which had the happy result of delaying China's nuclear weapons programme by over a decade. China's single 'Xia' class SSBN is equipped with 12 missiles that are credited with a range of 2700 kilometres and carry a single two-megaton warhead.

The end of the Cold War will probably lead to a decline in Western SSBN construction. China is still three submarines short of a minimum force, so further 'Xia' submarines may follow. However, nuclear missiles cannot be 'uninvented'. They remain the most reliable means of maintaining a nuclear deterrent and they will continue to stalk the world's oceans for the foreseeable future.

HMS Vanguard is the first of four British missile submarines ordered to replace the ageing 'Resolution' class boats.

USS CALIFORNIA

The US Navy's most important warships are its nuclear-powered aircraft carriers. They form the centrepiece of Task Forces operating all over the world, their powerful air groups enabling them to dominate vast stretches of ocean. However, these mighty vessels are vulnerable. A few hits from modern surface-to-surface missiles could disable them for many hours – and modern naval battles are often decided in minutes, or even seconds.

The US Navy has several classes of ships specifically built to support the aircraft carriers and protect them against attack. Stand-off missiles and over-the-horizon targeting could enable even minor navies to mount a deadly missile attack on a US carrier. Such a threat must be met as far away from the carrier as possible and, during the 1950s, the Navy introduced several long-range surface-to-air missiles. The launchers and associated radar systems were too bulky to be fitted to the numerous wartime vessels then in service and a new generation of cruisers had to be built to operate them. The 'Leahy' and 'Belknap' class cruisers were commissioned during the early 1960s: displacing 4,500 and 6,500 tons, respectively, they were de-

The nuclear-powered cruisers were designed to provide long-range SAM cover. They normally operate in pairs but, here, both 'California' class cruisers lead four 'Virginia' class.

USS California is an 8,700-ton nuclear-powered cruiser commissioned in 1974. Her two pressurised water-cooled reactors develop 70,000 shp, driving the ship at over 30 knots.

signed to keep up with the carriers in the heavy seas of the North Atlantic.

The US Navy had already put a nuclear-powered cruiser into service. USS *Long Beach* was a radical departure from traditional cruisers and not just because of her two Westinghouse pressurised water-cooled nuclear re-

actors. *Long Beach* carried no guns – just Talos and Terrier surface-to-air missiles. Although two 5-inch guns were added after the missiles repeatedly missed their target during a practice shoot attended by President Kennedy, *Long Beach* demonstrated the SAMs' capabilities in Vietnam. Cruis-

ing off the North Vietnamese coast, the cruiser shot down several MiG fighters at ranges of over 100 kilometres.

The nine 'Belknap' class cruisers commissioned between 1964 and 1967 were also nuclear-powered and built to provide the carriers with long-range air defence. They too were deployed off the Vietnamese coast as US Navy aircraft attacked targets in Vietnam. USS *Sterett* achieved a technological 'first' on 19 April 1972,

shooting down an incoming SS-N-2 'Styx' surface-to-surface missile with a Terrier SAM. The same cruiser also brought down two MiG fighter-bombers at ranges of 9 and 27 kilometres, respectively. Three months later the North Vietnamese tried again, sending five MiGs out to attack the fleet only to run into Terrier missiles from USS *Biddle*. Two MiGs were shot down before getting within 30 kilometres of the ships and the others aborted their attack.

USS *California* and USS *South Carolina* were laid down in 1970, the first American warships to be fitted with improved D2G reactors that have three times the core life of those in earlier cruisers. These two warships were to have been the first of five guided-missile 'destroyers', but the others were cancelled in favour of the four similar 'Virginia' class.

Naval terminology has yet to catch up with the missile age and warships are still described using the accepted classifications of World War II. Today the 'California' class are rated as cruisers, although these cruiser-sized warships were officially classed as frigates, and then guided-missile destroyers until 1975.

The 'Californias' main weapon is the Standard SM-1MR surface-to-air missile, which has a maximum range

Aboard USS *California*

USS *California*'s main weapon remains the Standard surface-to-air missile, but the addition of Harpoon has increased the ship's capability to deal with surface threats as well as air attack.

Standard missile launchers
California has single Mk 13 Mod 7 launchers forward and aft, each with 40 Standard missiles. These have a range of 46 kilometres against aircraft, but only very limited capability against surface ships. Hence the addition of Harpoon missiles during the 1980s.

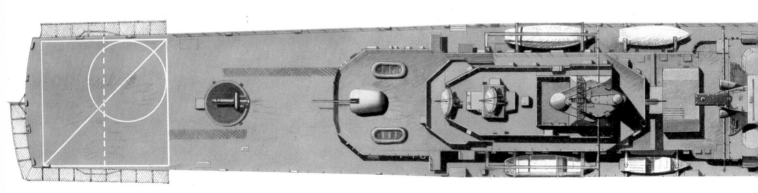

of 46 kilometres. Each cruiser carries 80 Standard missiles that are fired from two single-rail launchers. Two quadruple Harpoon surface-to-surface missile launchers (no reloads) provide an anti-surface capability. The Standard missiles and the pair of 5-inch dual-purpose guns are controlled by the Mk 86 weapon control system.

Surface action

In addition to air defence, the Mk 86 can handle surface action and shore bombardment, together with subsidiary functions of navigation, command and control. It can interchange data with the majority of US naval ships through the Naval Tactical Data System (NTDS) and through the Integrated Tactical Amphibious Warfare Data System (ITAWDS). Computer back-up is powerful enough to keep track of 120 simultaneous returns; separate aerial or surface targets can be engaged by each of the ship's weapon systems, which are able to switch to a new threat virtually instantaneously.

The Mk 86 system requires a variety of sensors. At the foremast head is the antenna of the SPS-48, tasked with three-dimensional air search. Beneath the radome on the main mast is the rapidly-rotating antenna of the SPQ-7, which gives fast response against low-flying targets out to about 40 kilometres. The set is frequency-agile to counter ECM and other interference.

The SPS-40 search antenna at the main-mast head can also give a measure of defence against the low flyer. Targets designated by these sets are acquired and tracked by the SPG-60 dish over the bridge. Abaft this antenna are the four SPG-51D dishes, a pair of which are associated with each Standard launcher. With tracking handled elsewhere, each pair of antennas can illuminate the target with great precision, one emitting a continuous energy wave and the other receiving the reflected signal, measuring the relative target speed by its Doppler shift.

With a 'California' class ship tucked in on either beam, a carrier is both well protected and performance-enhanced through the cruisers' additional sensors and data-links. While highly efficient in such a 'goalkeeper' role, knocking down threats to the carrier, the ships lacked, until recently, any additional capacity to guarantee the elimination of threats to themselves, typically a 'leaker' that penetrated the main defence ring. The Standard MR is viewed, it would seem, as suitable to double for a point-defence system, which is not fitted (even it it were, the standard Sea Sparrow is badly overdue for replacement). Two Vulcan-Phalanx CIWS mountings are fitted. Even these fast-response and autonomous weapons have shortcomings. Current thinking favours a 25-mm or 30-mm CIWS that has the range and stopping power to

explode or disintegrate a hardened SSM at a safe range; even with superdense penetrators, the Vulcan's 20-mm projectiles are apparently on the puny side even if the extra rate of fire is taken into account. Current Vulcan mountings cannot elevate to an

FMC 5-in Mk45 guns
The two turrets house single 5-inch (127-mm) guns which can engage surface targets at a maximum range of 23 kilometres and aircraft within 15 kilometres. For close defence against aircraft or missiles, the 'Californias' rely on their Phalanx six-barrelled 20-mm cannon.

ASROC launcher
For defence against submarines the 'California' class are armed with a Mk 16 octuple launcher for ASROC anti-submarine missiles. This delivers a Mk 46 Neartip anti-submarine homing torpedo to a maximum range of 10 kilometres.

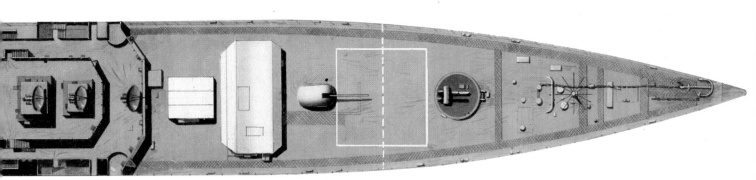

USS **Long Beach** *was the first US surface warship to have nuclear power. Fitted with Talos SAMs, she shot down several MiGs over North Vietnam at ranges of over 120 kilometres.*

ship lacks her own helicopters and, at the speeds at which she would normally operate, her hull sonars would be degraded in performance.

On the 'Virginia' class the overall design was improved by adopting twin-arm Mk 26 launchers, which can handle both Standard and ASROC rounds. With the dedicated ASROC launcher and its space-consuming re-load facility removed, the superstructure could be sited farther forward. Even with this addition, the 'Virginia' class ships are based on a hull some 3.35 metres shorter than that of the 'California' class. Although this reduction graphically illustrates an improved layout, it is hardly defensible hydrodynamically and will prove a positive shortcoming when it comes to modernisation.

Below: USS **Sterett** *was the first warship to shoot down an incoming surface-to-surface missile.*

angle sufficiently high to combat the steeply-diving missiles, even though the ship's own Harpoon SSMs can use this attack technique.

Nuclear submarines pose one of the gravest threats to a carrier, and even an escort of the size of the *California*

Below: Commissioned in 1962, USS **Bainbridge** *was the second nuclear-powered cruiser. She has served in the Pacific Fleet for over 30 years.*

carries an ASROC launcher. Though this is a stand-off weapon, effective out to about 9 kilometres, targeting information would usually need to be acquired from other platforms as the

USS *TARAWA* ONE-SHIP INVASION FLEET

Its massive slab sides tower out of the water as though a huge building had somehow floated out to sea. It looks industrial, rather than warlike. But the USS Tarawa is indeed a warship, and an extremely valuable one. It is one of the most capable amphibious assault ships ever built.

Amphibious warfare has been around in one form or another for a long time. Julius Caesar transported an army across what was to become the English Channel in the 1st century BC. The English transported armies the other way 1400 years later, during the 100 Years War with France. But the art of conducting amphibious assaults reached its peak in the Pacific campaigns of World War II and in the massive invasion of Normandy in June 1944.

New designs of ship and landing

Above: A US Marine Corps CH-46 Sea Knight assault helicopter lifts off the deck of a 'Tarawa' class general-purpose assault ship. Amongst their many capabilities, the 'Tarawas' are the world's largest helicopter-carriers.

The huge slab sides of the USS Tarawa tower out of the Caribbean as a US Marine Corps Harrier hovers before landing on the broad flight deck. The ship is flooded down aft, with its cavernous docking well ready to receive landing craft.

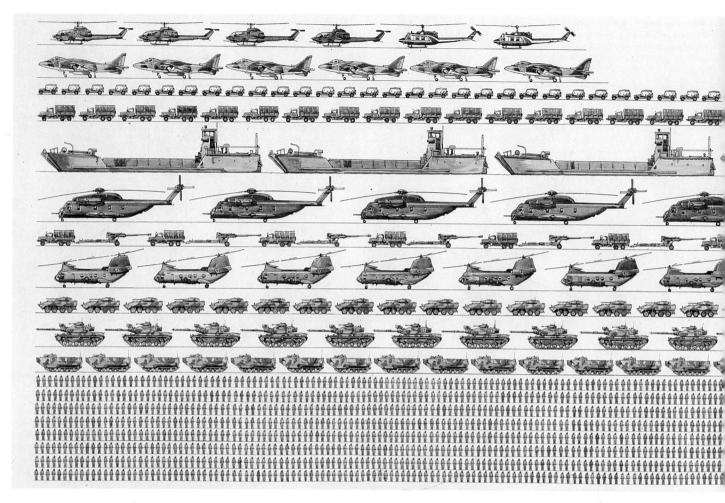

craft were evolved by the Royal and US Navies for these operations. Some were specialised cargo ships, with landing craft carried in place of lifeboats. Some had docking wells to allow large landing craft to be carried. Others were designed to carry tanks and heavy equipment. After the war the advent of the helicopter saw the appearance of dedicated helicopter carriers, often converted from light aircraft carriers. These saw action for the first time during the 1956 Anglo-French landings at Suez.

The Americans possess the world's largest amphibious fleet by far. Specialised amphibious vessels are expensive, and with the decline of the European Colonial empires, only the United States needed that kind of long-range power projection capability. More to the point, only the US Navy could afford to maintain such a massive force, let alone spend the money to acquire new, better assault vessels.

New designs of amphibious vessel have appeared since the 1950s. These are usually upgraded variants of the types developed in the 1940s. The 'Tarawas', which were commissioned in the 1970s, are something else. They are big. Twice the displacement of Britain's 'Invincible' class light carriers at nearly 40,000 tons, they are known as LHAs, or general-purpose amphibious warfare ships. Their size is dictated by the need to be able to perform the functions of several vessels at once.

One 'Tarawa' combines the aviation capacity of the LPH helicopter carrier, the docking well of the LPD amphibious transport dock, the command and control facilities of the LCC amphibious flagship and the cargo-carrying capacity of the LKA attack transport. Capable of carrying the 1,900 men of a reinforced marine battalion, together with all their equipment, the 'Tarawas' are virtually complete landing forces contained within a single ship.

An amphibious task force com-

Assault ships like the Tarawa have all the grace of a floating brick. The 'Tarawas' and the succeeding 'Wasps' are nevertheless far and away the most capable amphibious warfare vessels the world has ever seen.

Warload of the *Tarawa*

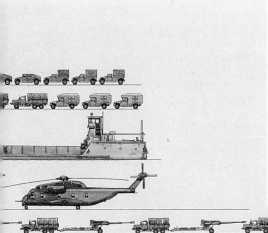

The USS *Tarawa* has a huge warload. The ship's main task is to land the 1,900 men of a reinforced Marine Amphibious Battalion, together with their equipment. Typically, this could incorporate attack, assault, transport, and utility helicopters. Six to eight Harriers can be carried to add versatility to the air complement. Numerous light and medium trucks, mostly jeeps, 'Hummers' and standard 2½-ton vehicles are also part of the package. Fighting vehicles in the warload can include a battery of 155-mm howitzers, a Light Armored Vehicle company, a tank company, and a company of amphibious assault vehicles, which is capable of landing 600 troops at one time. Lighter items are air-portable beneath CH-53E Super Stallion helicopters, while the heavier items are ferried ashore in one of the vessel's four 200-ton utility landing craft.

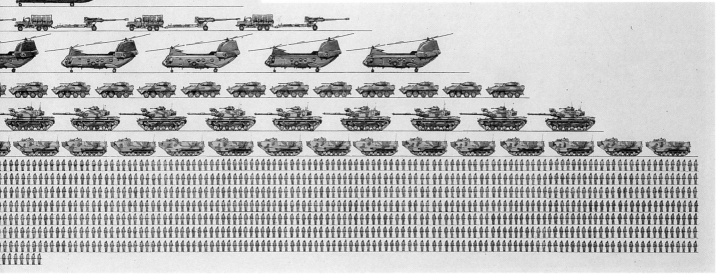

mander needs to maintain control of his forces at all times. The 'Tarawa' class is fitted with comprehensive command and control equipment and satellite communications gear. Known as the Tactical Amphibious Warfare Data System, it enables a commander to make the most effective use of an amphibious group's aircraft, weapons, landing craft and sensors.

Landing troops

The LHA's main function is to get troops ashore in the shortest possible time. Troops and lighter items are air-portable beneath CH-53 helicopters, while assault troops can drive their AAV-7 amphibious assault vehicles straight from their garage decks into the water and onto the beach. Heavier pieces of equipment such as tanks are ferried ashore in one of the vessel's four 200-ton LCU utility landing craft. These can use the floodable docking well at the stern of the ship to load troops and heavy equipment like tanks. Measuring more than 80 metres in length and nearly 24 metres wide, the docking well can hold up to 20 of

the smaller 60-ton LCMs in place of the four LCUs. 'Tarawas' can also operate with air-cushion landing craft, but not as well as the later 'Wasp' class vessels, which have docking wells designed specifically to accommodate three LCAC air-cushion craft.

Helicopters have been an important part of the US Navy's amphibious capability since the 1950s, when the US Marines first developed the concept of 'vertical envelopment' – avoiding enemy defences by going over them. The five members of the 'Tarawa' class each have hangars for up to 30 CH-46 Sea Knight or 19 CH-53 Sea Stallion helicopters. Another

The USS Nassau (LHA-4) hosted AV-8B Harriers from Marine Corps Squadron 331 during the 1991 Gulf War. In the process the Nassau became the first US Navy assault ship to conduct fixed-wing combat operations.

dozen CH-46s can also be carried on deck, however. A typical helicopter group might consist of 12 CH-46s, six CH-53s, four AH-1T SeaCobra gunship helicopters, plus two UH-1 'Huey' utility helicopters.

Fixed-wing aircraft

The 'Tarawas' are not just helicopter carriers, however; they can also operate fixed-wing aircraft. The acquisition of the British-designed Harrier V/STOL fighter-bomber has given the Marines and the US Navy's amphibious fleet real, instantly available air power. Currently, the Marine Corps operates the vastly improved AV-8B Harrier II. The normal Harrier complement is six to eight aircraft, but more can be carried at the expense of helicopters. Harriers from VMA-331 operated from the USS *Nassau* (LHA-4) during the Gulf War, while the USS *Tarawa* was equipped with an all-helicopter group.

Tarawa has a huge warload. Apart from the utility, transport and attack helicopters and the AV-8B Harrier fighters, the vessel has over 3000 square metres of parking space on its vehicle decks. This can accommodate 160 vehicles. A typical load might include a tank company with M60 or M1 Main Battle Tanks, a light armoured vehicle company, a battery of 155-mm howitzers, and trucks. In addition, an assault amphibian company with its 40 AAV-7 amphibious assault vehicles can be carried. There are 3300 cubic metres of palletised storage space, and tankage for 40000 litres of vehicle fuel and 400000 litres of aviation fuel.

The LHA has accommodation for the 1,900 men of a reinforced Marine Amphibious Battalion. They have a 500 square metre acclimatisation and training room so that they can exercise in a controlled environment. A 300-bed medical unit is also carried. This is a fully-equipped hospital with operating theatres, X-ray equipment, laboratories, a pharmacy, a dentistry facility and full medical stores.

Redesigned 'Tarawas'

The 'Tarawa' class was intended to number nine vessels, but the rundown of US forces after the Vietnam War, together with tight budgets, meant that only five were built. The 'Tarawas' were the largest assault ships in the world, until the entry into service of the USS *Wasp*, the first of a new class of assault ship designated LHD. Marginally larger than the preceding LHAs, the 'Wasps' are basically redesigned 'Tarawas', having the same general configuration but with more efficient use of space and better survivability.

USS **Nassau** *cruises the Gulf in formation with the new 'Whidbey Island' class dock landing ship* **Gunston Hall.** **Nassau** *formed the heart of the US Navy's Amphibious Group Two in the Persian Gulf. The dozen vessels carried the men and equipment of the 4th Marine Expeditionary Brigade.*

Right: The USS **Wasp** *is the first of its class. Designed to replace the 'Iwo Jima' class of helicopter carrier, the 'Wasps' are similar to but marginally larger than the 'Tarawas'. They have a smaller island structure, and command and control spaces have been moved from the island into the hull to reduce the possibility of a 'cheap kill'.*

The Falklands war demonstrated the wisdom of continuing to arm warships with a gun. The Type 21 'Amazon' class frigates carried out several bombardments of Argentine positions on the islands.

VICKERS 4.5-in GUN

The Royal Navy flirted briefly with the idea of the all-missile warship in the 1970s, when the original Type 22 'Broadsword' class frigates were designed. But experience in the Falklands showed that the medium naval gun still had a place, even on the most advanced of fighting vessels.

Shore bombardment in support of amphibious landings was the requirement in the South Atlantic, but when allied to advanced fire-control systems the medium gun is also an effective counter to anti-ship missiles, its large explosive charge making up for a relatively slow rate of fire. Batch 3 examples of the Type 22 design and the latest Type 23 class of frigates have included a 4.5-in gun

Guns are also effective weapons in

the kind of warfare which has developed in the 1990s, where naval blockade has become an important weapon in the arsenal of peacekeeping. It is rather difficult, and very costly, to warn a ship trying to get into Iraq or carrying munitions for Serbia when the only weapons you have available are surface-to-surface missiles. Anti-ship missiles are designed to hit the target, not near miss. It is also very expensive, with weapons like Harpoon or Exocet costing tens or even hundreds of thousands of pounds. It is far simpler to fire a couple of shots across the bows.

The Royal Navy has been using 4.5-in (114-mm) calibre weapons as its standard medium calibre since before the end of World War II. The Mk 6 gun was most widely used in destroyers

and frigates, most recently being carried in a twin turret aboard 'Leander' class frigates and 'County' class missile destroyers. The last Mk 6-armed vessels were withdrawn from service at the end of the 1980s, although a number remain in service with other navies.

In the mid-1960s, the Royal Armament Research and Development Establishment began to design a fully automatic weapon to replace the Mk 6. Based on the British Army's Abbot 105-mm self-propelled gun, the radar-controlled 4.5-in Gun Mk 8 entered service in the early 1970s with the commissioning of the Type 82 and Type 42 class destroyers.

The Mk 8 is a 55-cal weapon fitted with a muzzle brake and a fume extractor. The gun mounting itself was designed by Vickers and features a hardened glass-reinforced plastic gun shield and a fully automatic hydraulic loading system that has been kept as simple as possible. Although the gun has a relatively slow rate of fire, it is very quick to get into action. It only takes 10 seconds to power the system up from complete shut down to fire the first round. The barrel is air-cooled and can maintain a high rate of fire for long periods – 90 rounds in 7.5 minutes – without appreciable decrease in accuracy.

Ready-use ammunition

Ready-use ammunition is stored vertically on a carousel below deck, with 16 rounds of one-piece ammunition being available. The main ammunition type carried is dual-purpose high explosive, equally suitable for air or surface action. The HE round's fuse is set electronically as the round is being rammed into the breech, and the gun's operators can choose between impact, close proximity, distant proximity and delayed action modes.

Four other types of fixed ammunition can be fired: chaff, surface practice, anti-aircraft practice and illumination. Two spaces on the ready-use carousel are normally reserved for illuminating rounds.

Above: **HMS** Arrow *provided naval gunfire support for 2 Para at Goose Green, but the gun broke down early in the battle, leaving the soldiers in the lurch.*

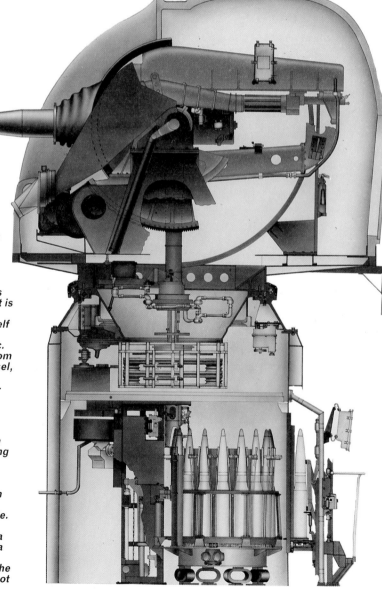

In the late 1980s, the Royal Navy initiated plans to increase the effectiveness of the Mk 8, specifically with the development of an extended range base-bleed round that will increase the gun's effective range from 23 to 27 kilometres, but this has yet to be ordered.

The Mk 8 is currently in service with the Royal Navy and with a number of other navies, including those of Argentina, Brazil, Iran, Libya, Nigeria and Thailand. It saw extensive use in the South Atlantic in 1982, mainly in shore-bombardment and close-support roles, but it is also credited with the sinking of an Argentine supply ship when HMS *Alacrity* sank the *Isla de los Estados*.

Right: The Royal Navy's 4.5-in gun is radar-directed and capable of engaging enemy aircraft as well as surface targets. It is fully automated and the turret itself is made of glass-reinforced plastic. The gun is fed from a 16-round carousel, which can be loaded with high-explosive or illumination rounds. The gun also fires chaff shells, which can decoy an incoming anti-ship missile. Thanks to its automatic operation the gun can fire up to 25 rounds per minute. However, most ships only carry a single gun, so if a mechanical problem occurs the fire mission cannot continue.

Specification
Vickers Mk 8 Gun

Calibre: 114 mm
Weight: total 25.75 tonnes; rotating mass only 15.1 tonnes
Elevation: −10°/+55° at 40 degrees per sec
Training: +170° at 40 degrees per sec
Muzzle velocity: 870 m/sec
Projectile weight: 21 kg
Maximum rate of fire: 25 rounds per minute
Maximum effective range: surface 23 km, anti-aircraft 6 km